Asia in Transition

Volume 27

This book series, indexed in Scopus, is an initiative in conjunction with Springer under the auspices of the Universiti Brunei Darussalam – Institute of Asian Studies (http://ias.ubd.edu.bn/). It addresses the interplay of local, national, regional and global influences in Southeast, South and East Asia and the processes of translation and exchange across boundaries and borders. The series explores a variety of disciplinary and interdisciplinary perspectives.

Submission and Peer Review:

Proposal submissions are to be sent to the Series Editor, Dr Bruno Jetin: bruno.jetin@ubd.edu.bn and Springer Publishing Editor Alex Westcott Campbell: alexandra.campbell@springernature.com using the Book Proposal Form available in the sidebar. All proposals will undergo peer review by the editorial board members. If accepted, the final manuscript will be peer reviewed internally by the editorial board as well as externally (single blind) by Springer ahead of acceptance and publication.

Himanshu Upadhyaya

Critical Insights on Colonial Modes of Seeing Cattle in India (1850–1980)

Tracing the Pre-history of Green and White Revolutions

INSTITUTE OF ASIAN STUDIES

Springer

Himanshu Upadhyaya
Azim Premji University
Bengaluru, Karnataka, India

ISSN 2364-8252 ISSN 2364-8260 (electronic)
Asia in Transition
ISBN 978-981-97-1559-6 ISBN 978-981-97-1560-2 (eBook)
https://doi.org/10.1007/978-981-97-1560-2

This Springer imprint is published by the registered company Springer Nature Singapore Pte Ltd.
The registered company address is: 152 Beach Road, #21-01/04 Gateway East, Singapore 189721, Singapore

Contents

List of Tables

Chapter 1
Introduction

Historically bovine animals (cattle and buffaloes) rearing has been critically linked to livelihood strategies of peasants in the agro-pastoral landscape Indian subcontinent. Literature from pre-colonial and early colonial periods underlines mutually reinforcing 'symbiotic' relationship between crop cultivation and cattle keeping in Indian farming. These authors also speak about an interdependence between settled cultivator and nomadic cattle breeders or transhumant pastoral people.[1] However, such an interdependence was turning out to be fraught with extension of cultivation and more and more land being brought under cultivation in the late nineteenth century and early twentieth century in India.

This book attempts to offer critical insights on colonial modes of seeing cattle in India. By engaging with fresh material from archival records that invites readers to consider bovine animals in India as having subjecthood, this book reviews the impacts of the repeated discursive acts of giving them the social, cultural and material life. In doing so, this book strives to answer the questions regarding the attitudes towards bovine animals that were articulated by colonial experts as well as peasants and Indian social and political leaders. Reading of archival records on bovine animals in India carrying the viewpoints of officials from the British Raj—which oscillates a great deal between an exalting veneration of "the cow and the working bullock as bearing the whole structure of Indian agriculture" (Hope 1937) and "breeding bulls as standing religious menace" (Woelcker 1893)—thus gets presented in this book as an effort to make sense of late colonial state and its Science and Technology institutions. This book attempts to do that by paying close attention to how such representations ended up in turn guiding the perception of Indian peasants and their allies towards bovine animals. While agrarian history of India has paid attention to what happened to crops, during the colonial and late colonial period, what happened to the bovine animals in late colonial India has remained an understudied aspect.

Probably the reason for the lack of engagement with the social, economic and environmental history of cattle in India can be traced in the tendency to engage with two ideas that started to frame the policy discourses around bovine animals in India.

H. Upadhyaya, *Critical Insights on Colonial Modes of Seeing Cattle in India (1850–1980)*, Asia in Transition 27, https://doi.org/10.1007/978-981-97-1560-2_1

To be precise, the notion of 'sacred cow' and hence 'surplus cattle' in India. The first notion, 'sacred cow' has made historians focus on the contours of meanings attached to this domesticated animal ranging between the great veneration accorded to her to a mischievous enrolment of her symbolic and iconic value in political and communal mobilizations. At the same time, the second notion, 'surplus cattle' has been reflected on for several decades in writings by livestock economists, anthropologists and other academicians. Much of these writings surprisingly do not trace the relationship amongst cattle, cattle keepers and agrarian as well as agro-pastoral landscapes. While environmental history writings in India remained focused on imperial administration of forested landscapes a great deal alongside remaining preoccupied with discussions on the arrival of perennial irrigation and transformations it brought about on agrarian/agro-pastoral landscapes, there is comparatively less engagement with what transformations were prompted by the arrival of veterinary medicine, veterinary education and along with it the notion of 'scientific livestock management/ livestock improvement'.

This book attempts to trace contours of mutually reinforcing symbiotic relationship between crop cultivation and cattle rearing by reading against the grain of several official accounts from the late colonial period in India. In part, we witness an acknowledgement of a thick set of relationship and inter-dependence between a nomadic cattle breeder and a settled cultivator, between land-livestock-agriculture in the writings of Sir Albert Howard (1873–1947), who advocated some of the most sophisticated ideas on integration, holism and mixed farming in an era when agricultural research was marked by increasing specialization and compartmentalization. This book's major contribution to academic scholarship is rooted in its deep engagement with dissenting voices articulating their vision and concerns about the relationality between crop cultivation and cattle rearing, while being part of Science and Technology Institutions in late colonial State (i.e. British India). In doing so, it widens the academic scholarship within the discipline of History of Science, Technology, Environment and Medicine (STEM) in India and builds on the early contributions such as *Beastly Encounters of State* by Saurabh Mishra, *Meat, Mercy and Morality: Animals and Humanitarianism in Colonial Bengal* (1850–1920) by Samiparna Samanta.

This book brings hitherto unknown facets around the pre-history of India's Dairy Revolution (white revolution or sometimes known by its official name Operation Flood) by discussing in details how colonial state was discussing the efforts and intervention to lead India on the path of organized dairying and modernization of cattle rearing and dairy farming. By presenting accounts of early milk cooperatives that were highly successful (e.g. Calcutta Milk Union 1917–1940s), this book for the first time opens up a detailed pre-history of Kheda District Cooperative Milk Union and Varghese Kurien-led Operation Flood (white revolution).

1.1 An Outline of How Academic Scholarship Has Engaged with Bovine Animals So Far

Even as environmental history had announced its entry amongst modes of accessing past through the art of historical research in 1980s in India, reviewing a new book by an Irish economist, Raymond Crotty, *Cattle Economics and Development*; Harold K. Schneider made a remark in a somewhat challenging and inviting tone, "the special role played by livestock and their special relation to agricultural operations in socio-economic systems throughout the Third World has been neglected long enough".[2] Raymond Crotty had also inserted a suggestion in his book that "there may be a scope for an international institution to research, teach and advise on exploitation of the world's pastoral resources". When Crotty made this suggestion calling for an international institute, there had existed just one institute which had come up in July 1974 due to collaboration between Ethiopian government and World Bank. This institute—International Livestock Centre for Africa (ILCA)—ran "research programmes for major ecological zones of Africa involving research in animal sciences, environmental sciences and socio-economics" for about twenty years before merging on 31st December 1994, with International Laboratory for Research on Animal Diseases (IRAD)—a Kenya-based institute which was established in September 1973 as a result of collaboration between Government of Kenya and The Rockefeller Foundation—to form International Livestock Research Institute.[3]

Around late 1980s, a well-known agricultural economist, C. H. Hanumantha Rao acknowledged candidly that "agricultural economists in India have been interested essentially in the economics of crop production and comparatively less in livestock economics".[4] He added qualifying his remarks that while there were a few studies that explored the impacts of agricultural development on ecology, these studies had not explored trends and patterns of crop production and livestock economy in depth.[5] While many economists did (and still do) engage with issues around livestock development in India in general and contribution of livestock keeping in GDP in specific, within the discipline of livestock economics there has been long wielding discussion and debate around 'the sacred cow complex', where the overwhelming concern revolves around the notion 'surplus cattle'.[6] However, when one engages with this prolonged surplus cattle debate amongst economists, anthropologists and social scientists, one gets a feeling that the debate gets carried out without reflexivity on the discursive impact.[7] It is also surprising that even with a substantial population of bovine animals, small ruminants and fowls that India accounts for, there are no professional organizations around the academic discipline, 'livestock economics' and no academic journals in India devoted to 'livestock economics'.[8]

Of late, there have been some efforts to address this lack of academic scholarship and professional work. Thus, we see that within the civil society space, there is now a well-recognized consortium of organizations that work in the livestock and Natural Resources Management (NRM) sectors, namely Rainfed Livestock Network.[9] However, within the academic realm a lot of writings from Livestock

Economics lens still have to rely on journals such as *Indian Journal of Agricultural Economics*, run by Indian Society of Agricultural Economics or *Review of Agrarian Studies*, run by the Foundation for Agrarian Studies since there is no exclusive academic journal devoted to the cause of Livestock Economics in India.[10] Nitya Ghotge, a scholar-activist has written about the role livestock plays in sustaining livelihoods in India in her book, *Livestock and Livelihoods: The Indian Context*.

In similar vein, "the post eighties emergence of environmental histories put the focus squarely on forest and water management", repeatedly drawing the attention of the readers on the devastating impacts of 'scientific forestry' on India's agrarian landscapes. However, this early academic scholarship within environmental history had offered very little on what were the impacts of the imperial ideas on 'livestock improvement' and bringing the veterinary science and medicine to India within the overall paradigm of imperial science and technological institutions.[11] So, when this gap in the environmental history scholarship got addressed by D. Satya Laxman's 2004 book, *Ecology, Colonialism and Cattle: Central India in the Nineteenth Century*, a reviewer remarked that "In the case of Satya's book, there is no doubt that he has selected an issue that while being absolutely fundamental to an agricultural society like India, has surprisingly been neglected in existing historical accounts".[12]

Essays published in the special section titled 'Cows, Buffalo, Bullock: Bovine Politics in South Asia' (in *South Asia: Journal of South Asian Studies*, Vol 42, No 6, pp. 1095–1209) has contributed to disruption of three powerful framing binaries that has animated the academic discourse around the politics of cow protection in India: religion/economy, legality/illegality and cow-lover/cow-killer.[13] The essays pay careful attention to "the material, symbolic and affective work that bovine bodies do, and have done, in South Asia". Contemporary academic scholarship now recognizes that "the politics of cow protection is evidently entangled with practices of caste labour, with farming economies, dietary cultures and situated human-animal relationships" (Adcock & Govindrajan 2019: 1096).

The dependence of agriculture on cattle in India was so crucial that Victor Hope, Governor General of British India (1936–1943) had stated in his opening speech at the Cattle Conference in 1937, eloquently:

> The cow and the working bullock bear on their patient back the whole structure of Indian agriculture. In actual life of the people and in the rural economy of the country, the relative importance of bovine animals is indeed much more marked than the figures of the total number of cattle suggest...In most parts of the world cattle are valued for food and milk; in India their primary purpose is draught for the plough or the cart...Without the ox, no cultivation would be possible, without the ox, no produce could be transported.[14]

Saurabh Mishra (2015: 02) exclaims that given the centrality of cattle in the lives of peasants, "it is surprising to note that, though there is an extensive body of historical work on agrarian processes and structures, the intrinsic importance of cattle in the life of Indian peasants have received comparatively scant attention from historians".[15] Mishra argues that while "cursory references contained in certain works do bring livestock's centrality to the fore", there is very little effort "devoted towards exploring this link between cattle and the agrarian landscape any further".

Extending his argument, Mishra (2015: 03) states that "the question of cattle has been ignored not just by scholars working on agrarian conditions, but also by historians of medicine in India". Mishra finds it quite surprising that historians of medicine "have completely sidestepped the question of veterinary health in South Asia in spite of huge cattle mortalities that were caused by epizootics".[16] Mishra considers that the tide has begun to turn in last couple of decades with several remarkable works on the history of veterinary medicine being published, starting with first full-length article on veterinary history being carried by reputable journal *Social History of Medicine* in 2002.[17]

Within the realm of social anthropology and anthropology, scholarly engagement with Nomadic groups in South Asia had surfaced only in the 1970s. However, "with a few notable exceptions, even this interest was largely limited to expounding the 'problems' these 'backward' communities created for the state in its developmental aims and in finding solutions to these", felt Aparna Rao and Michael Casimir while writing an introduction for an edited volume, *Nomadism in South Asia* (2004) that claimed to be the "first effort to present comprehensive information and academically informed discussions on the nomadic peoples in South Asia".[18]

In their study of crop, livestock and livelihood interactions in Africa, science policy researchers identify the emergence of mixed farming paradigm owing its inspiration from 'the agricultural revolution' in Britain in the eighteenth and nineteenth centuries, when a particular package of scientific techniques started to be employed. Campbell and Overton (1993) unpacks this mixed farming paradigm by revisiting six centuries of Northfolk farming (circa 1250 to circa 1850) and arguing that it included technical innovations such as manure intensive husbandry, the use of legumes, a reduction of fallows and the integration and mutual development of arable and pastoral husbandry. While such an emergence of mixed farming paradigm might have evolved and found favours from soil scientists and others from 1843 onwards with the establishment of Rothamstead trials, the kinds of agricultural and veterinary interventions in India had overwhelming preoccupation with plant pathology and cattle disease concerns. Rather than having a keen eye and ear for acknowledging the knowledge and agro-pastoral practices of Indian peasants, experts at Pusa and Muktesar had built up such an apparatus of agricultural and bacteriological research that Sir Albert Howard and Gabrielle Howard had to decide to move away from, what to them appeared as 'an artificial subdivisions of science'. Howards then chose Institute of Plant Industries at Indore to be a site wherein to try out and test their ideas on integration.

While writing about the genesis of the Institute of Plant Industries at Indore, Sir Albert Howard stated how during the scope of working on a range of issues at Imperial Agricultural Research Institute at Pusa (Bihar) for almost 20 years (1905–1924), "it became increasingly evident that the organization of an agricultural research institute, on the basis of practical agriculture on one hand and on the separate sciences on the other, was by no means the ideal arrangement". In his 1929 book, *The Application of Science to Crop Production*, coauthored with his wife Gabreille L. C. Howard, he didn't mince words and termed the organization of agricultural research in India as obsolete and since "*it is not easy to change any form of organization from within, this involved the foundation of a new institute of crops, at which the development of plant could be studied as a biological whole and not piecemeal*".[19] (emphasis added).

Similarly, in their book, *The Development of Indian Agriculture*, Sir Albert and Gabrielle Howard (1929: 54–55) alerted their contemporaries to guard against fragmentation in the following words:

> In the earlier years of the present century, the cattle problems of India were approached indirectly from pathological side. It was felt that the chief need was to preserve the existing cattle force by inoculating the animals against Rinderpest and other diseases...As in the case of crops of India, the cattle problem was first approached from the disease aspect. *It will be interesting to see how researchers on the livestock in the country develop in the future and whether the various sections on the subject – breeding, feeding, milk production and disease control – can be welded together into a single branch of agriculture, or whether the separate sub-divisions of the subject will persist...The evils which result when investigations follow the artificial sub-division of science rather than the problem itself are well known. One of the things to avoid in any future work on Indian cattle is fragmentation.*[20] (emphasis added)

Howards had stressed upon the need to apply an integrated approach while looking at crop cultivation and cattle rearing questions. They suggested that solving these questions required "a considerable knowledge of different sciences, a long experience in research work as well as firsthand acquaintance with agriculture itself" and cautioned that "if all this is not brought to bear simultaneously, the result can only be the accumulation of more and more data which may or may not be useful to some master-builder of the future".[21]

In the early 1930s, there had emerged a paradigm that had put forward similar proposals for keeping alive crop and livestock interdependence. Dr. Bernard Augustus Keen in his paper read at the Royal Institute of International Affairs on 1st March 1932, stressed upon the need to urgently dovetail the arable and animal husbandries into one 'mixed farming' system.[22] This pioneer British soil scientist had observed that "the cattle problem dominates the whole situation" and put forth his firm opinion that "there could be no rapid increase in crop yields until the mixed farming system is adopted". Following the Royal Commission on Agriculture report, Dr. B. A. Keen was seconded from Rothamsted for two years (1929–1931) to be Director of the Imperial Institute of Agricultural Research at Pusa in Bihar. An ardent warrior for academic freedom, Keen was unsparing in his criticism of the lethargy that had set in amongst research staff at Pusa. He had stated without mincing words that "At Pusa, very few new lines of work had been introduced for years past, and the existing methods were so badly out of date as to be not only useless but dangerous".

His opinion was sought on "the progress and future of the Institute of Plant Industries (Indore), which had been built up by Albert Howard and his wife," since "they had just left after seventeen energetic years" and "Keen rated the achievements of the Howards highly, as excellent scientific agriculture with successful extension work". Thus, he had recommended that Institute of Plant Industries (Indore) "be continued and expanded rather than be changed, to pursue the plans of Howards that Indore should become a centre for fundamental research in agricultural science". However, Keen was appointed to India during the years marked by onset of world economic depression and his plans for vigorous reconstruction of Pusa met with abrupt reduction of funds and he had to wage a valiant struggle to argue against the proposals by the Retrenchment Committee to close agricultural research. Although he was offered an extension in order "to remain and carry out his plans for reorganization, Keen found that the financial restrictions rendered progress impossible" and decided to return to Rothamsted in 1931. His paper that is cited here was basically his attempt to summarize his impressions of India. In this summary "Keen diagnosed India's agricultural problems as primarily due to waste of natural resources by uneconomic land-holding and fragmentation, by misuse of common land under excessive numbers of uneconomic livestock and by failure to organize adequate fodder even for the work animals". It was a natural corollary from his diagnosis that he would grant "animal nutrition and management top priority".[23]

The question that needs to be posed to understand the trajectory of 'agricultural progress' and 'livestock improvement' in India would be to try and understand why such valiant dissenters like Albert Howard and B. A. Keen remain relatively unknown and what their struggles against the High Modernist tendencies of colonial state entails. The mixed farming paradigm as advocated by Dr. B. A. Keen was emphatically reiterated by Dr. N. C. Wright in his advisory report to Imperial Council of Agricultural Research on the development of cattle and dairy industries in India in 1937. In his report, Wright stressed upon adopting mixed farming involving not only the utilization of all available manure, but also argued for putting aside a part of farm for cultivation of leguminous fodder crop, which has potential to contribute so markedly to soil fertility. Wright believed that doing so would provide an ideal source of food for cattle and specifically the Milch stock and hoped that such an approach would also help to cheapen milk production.

When one tries to look for signs over a curious question, if mixed farming paradigm had found purchase with researchers at Pusa and with the Imperial Council of Agricultural Research, one does come across scattered discussions about the mixed farming trials. There are also traces that Imperial Expert on Animal husbandry, Sir Arthur Olver had taken much interest in mixed farming experiments that were initiated at Pusa in mid 1930s.[24] Speaking at a joint discussion on 'Nutrition in relation to Human Beings, Farm Livestock and Crops'; Arthur Olver had stated that "the solution of this problem must lie to a large extent in the development of a well-balanced system of mixed farming in India".[25]

There are also signs that Gandhian thinkers and activists—for the concerns that this book touches upon, looking at the ideas and work of Satishchandra Dasgupta and J. C. Kumarappa is immensely helpful—who were committed to carry out the constructive programme in rural India had engaged with mixed farming paradigm. To cite an example, Satishchandra Dasgupta in his book, *The Cow in India*, written while serving a jail term at Alipore (Bengal), called upon peasants in India to adopt mixed farming paradigm:

> Growing food involves cattle rearing for the plough and for the milk. Let every householder be a cattle-and-plant farmer, organizing mixed farming for the villages. If these are accomplished, other things will come naturally to fit in with the overall scheme.... Every farmer is both an agricultural farmer and animal husbandry man and cattle breeder. Their

> differentiation was non-existent in India. A differentiation has sought to be made by Colonial Government between these functions of the cultivator...Every means should be taken to encourage mixed farming wherever the habit may be disappearing. Mixed farming will mean agricultural farming, dairy farming and cattle breeding. All these should go together.

This book attempts to understand how despite this initial advocacy of mixed farming paradigm and efforts to reintegrate the crop cultivation and cattle rearing, in mid-1960s development planning underwent a sudden paradigm shift. The moment of the arrival of 'miracle seeds' in India was accompanied by an ambience and milieu wherein 'intensification' had already become a buzzword. Intensification became the driving force in both the realm, crops and cattle. The key village programme and community development programme of yesteryears were suddenly given up and new interventions unfolded at the onset of Third Five Year Plan. A decade later when another revolution—white revolution—was unfolding, its planners saw in 'crossbred cow' a mirror image of what 'High Yielding Variety seeds' were to green revolution.

This book however begins the story of cattle in India not from green and white revolution decades. The narrative on cattle must begin with a historical depth that engages with the late colonial period when nomadic cattle breeders were being talked about as 'vanishing tribe'. The narrative must begin from the decade of 1860s, when cattle plague strikes at Calcutta agricultural exhibition and wakes the imperial powers to initiate investigations on contagious disease amongst cattle.

Chapter 2 engages with the long-drawn 'surplus cattle' discourse and how there is a need to not get bogged down by the conservative streak that continued to go on wielding around 'cow protection' rhetoric ad nauseam. However, at the same time one needs to probe the Malthusian roots of the surplus cattle discourse, understand the rewriting of agro-pastoral relations under the legal regime during late colonial period in British India that introduced grazing tax, poll tax at Cattle Fairs and restrictions on forest grazing. By situating the 'surplus cattle' discourse vis a vis the emergence of 'mixed farming' advocacy from 1920s onwards, this chapter underscores the arguments voiced by scholars who represented 'Discontents of Surplus Cattle Discourse'.

Chapter 3 presents the grave issue of cattle mortalities during famines as well as considerable loss of cattle due to contagious bovine diseases. This chapter engages with the insights that can be gleaned from the reports of Civil Veterinary Departments and hopes to add to the recent academic work on the history of veterinary medicine by scholars such as Saurabh Mishra and Samiparna Samanta.

Chapter 4 talks about that moment when traditional modes of cattle keeping, and cattle care were being confronted by the scientific knowledge that was communicated through pamphlets, demonstrations and exhibitions by experts from Veterinary and Agriculture Departments. Traditional knowledge of cattle keeping and cattle care that revolved around oral repertoire from elders and folk masters were fast being replaced by claims of the scientific cattle breeding and preventive inoculation against contagious disease amongst bovines such as rinderpest. This was also a moment when India was being imagined and represented as quintessentially an agrarian nation and peasants were being encouraged to apply science to bring more land under cultivation of cereals and cash crops. In the policy discussions, as the line that was repeated often was to encourage private enterprise for cattle breeding, it is crucial to understand what circumstances led to the traditional occupation of cattle breeding turning out to be 'not paying in accordance with the efforts made'. This chapter engages with discussion on the theme of cattle breeding amongst colonial officers by reviewing several official documents.

Chapter 5 engages in a more concrete manner with the emergence of urban milk markets and middle-class anxieties around the question of food adulteration. It explores how the privileging of food production function of cattle marked a departure from the role they had historically played as part of complex system of interactions between crop cultivation and animal husbandry at farm and local ecosystem level. It seeks to develop an argument on how with the increased urbanization and rising milk prices during the First World War, a specialized form of dairy farming had come into existence in the form of city milch stables. Tracing the change over from what earlier had been a 'pastoral mode of resource use' to a 'colonial mode of resource use', organized on the lines of what the Empire did at home, it points out how the emergence of urban milk markets formed a part of broader colonial policy of reorienting the practices of peasants in Bombay Presidency and elsewhere throughout the nineteenth and early twentieth centuries. It deals with emergence of dairying and varieties of modes of production and marketing of milk and milk products, in order to understand the changing contours of cattle keeping in late colonial India. It also brings to light lesser-known narratives on the history of early milk cooperatives in India and argues that colonial experts were positively inclined to advocate for cooperative mode of milk production and processing during discussions on dairy development policies. Even though there existed a thriving private trade in milk and milk products and a form of dairy development that was resource and capital intensive in the form of Military Dairy Farms, both imperial dairy expert W. Smith and imperial expert on animal husbandry Arthur Olver had always spoken passionately about organizing processing and distribution of milk and milk products on cooperative lines. Both these experts also argued against colonial government's decision on closing down of the research creamery at Anand! These narratives problematize the received wisdom on the history of dairy development in India, which almost always answers the question: 'what existed prior to Kheda District Cooperative Milk Producers' Union' with a phrase: 'Polson's monopoly'. Against such a backdrop, why policy makers in independent India did choose a bureaucratic and top-heavy model of City Milk Schemes over the task of organizing milk cooperatives within two- or three-tier structure also remains a moot question, which gets explored in the next chapter.

Chapter 6 tries to represent the early discussions on the question of Dairy Development and how the question of dairy education was taken up slowly by colonial administration from mid-1920s onwards. This chapter also engages with the emergence of technocracy and compartmentalization that Sir Albert Howard had warned against, by discussing threadbare the debate on who shall have the control over the cattle breeding activities: Agriculture Department officers or Veterinary Department Officers?

Chapter 7 looks at the writings of the dissenters such as J. C. Kumarappa and Satish Chandra Dasgupta who advocated integration of cattle and crop development agendas, instead of intensification during the decade of 1940s. It traces how in the aftermath of the Bengal Famine and the partition; the food policy of India underwent swift and abrupt changes. Elite leaders of the post-colonial nation-state were smitten by a desire to "catch up" with the Western model of economic development and pursued cattle development policies that had a tone of paternalism and attempted to sedentarise nomadic cattle breeders. This chapter reviews the synoptic history of the intensification strategies that were adopted during the 'green' revolution years in the realm of both crops and cattle development. It also traces the changes in livestock holding pattern of 'progressive farmers' inspired by the 'green' revolution agenda.

Chapter 8 reviews the emergence of 'Anand pattern' milk cooperatives and the discourse of 'replicating Anand' by asking whether the then Prime Minister L. B. Shastri was the first one to spontaneously propose the replication and call upon a young dairy technocrat or whether this speech became the launching pad of an ambition—to institionalise a national milk and milk product marketing board—that young Varghese Kurien had cultivated for about a decade then. By linking the discourses over dairy development from mid-1960s to 1970s with the advocacy for 'cooperativisation' of city milk supply right from 1920s, this chapter poses a question: was there a unique Anand pattern of cooperativisation of city milk supply or had KDCMPU unknowingly and unwittingly replicated the success achieved very early on by milk cooperative unions in other parts of the subcontinent such as Calcutta and Lucknow! It reviews the emergence of Operation Flood, the criticism of Operation Flood strategies that surfaced during mid 1980s and the responses from National Dairy Development Board to criticism.

The crop-cattle interrelations and transformations under the intensification strategies followed during 'green' and 'white' revolutions that are discussed in the foregoing chapters leave us with the following question. Has there been any revision in the dominant modes of thinking in such a post-revolutions regime when sustainability concerns have entered the discussion and pastoral perspectives are beginning to find their place in policy discussions? This chapter studies the bovine stock holding pattern and the changes that have been witnessed after Operation Flood by reviewing the data from Livestock Census as well as NSSO reports on Land and Livestock holdings.

The last and concluding chapter presents some of the findings from the study. It also presents the emergence of organic cultivation and organic animal husbandry practices that trace the inspiration in the writings of Sir Albert Howard and other dissenters.

Notes

1. As argued by Ratnagar (2004: 96–98) the interaction between the two realms identified by the terms *graam* (settled agrarian community) and *aranya* (the outside sphere) had always remained rather fluid and witnessed complementarities. For the evidence on which Ratnagar builds up her argument, see the special issue on 'Pastoralism in South Asia' in *Studies in History*, Vol 07, No 1, 1991.

2. See, Schneider, Harold K. (1983) 'Review of *Cattle, Economics and Development* by Raymond Crotty', *Economic Development and Cultural Change*, Vol 32, No 1, (October 1983), pp. 222–226.
3. See, https://uia.org/s/or/en/1100032753 and https://uia.org/s/or/en/1100068003. Last accessed February 14th, 2018.
4. Hanumantha Rao, C. H. (1988) 'Agricultural Development and Ecological Degradation: An Analytical Framework', *Economic and Political Weekly*, Vol 23, No 52/53, pp. A-142–A-146.
5. In the year 1987, annual congress of Indian Society of Agricultural Economists had for the first time scheduled a session on the theme 'Impact of Agricultural Development on Ecology and Environment'. For an account of this session, please see July–September 1987 issue of *Indian Journal of Agricultural Economics*.
6. Later in Chap. 2, this prolonged surplus cattle debate is discussed and around the different strands that engage with India's sacred cow complex, the narrative is structured.
7. By discursive impact, one attempts to bring to attention how a certain simplification gets staged that glosses over the possibility of references to livestock resources being invoked in Indian language/ by Indian scholars in pejorative manner. I grew up in a western state in India (Gujarat) and a linguistic community (Gujarati), known for its mercantile ethic. During my childhood, I repeatedly heard elders using the term 'Dhor DaakTar' in a pejorative sense. Similarly, responding to K. N. Raj's work, *Livestock in Agrarian Economic: Some Issues Concerning 'Sacred Cow' and 'Surplus Cattle'*, (Delhi School of Economics, University of Delhi, 1969), V. M. Dandekar used the title, 'Cow Dung Models' (*Economic and Political Weekly*, Vol 4, No 31, (02 August 1969). Saurabh Mishra refers to similar pejorative epithets—'horse-doctor' or 'farrier' in use for Veterinary practitioner in Britain and argues that one constant factor undermining veterinary medicine was its subordinate position vis a vis human medicine.
8. While in the higher education institutes, to some extent the discipline 'livestock economics' is engaged with—for example at Indian Veterinary Research Institute, Izzatnagar, Bareilly there is an M. Sc. In Veterinary Science program with specialization offered in Livestock Economics—there is an overall tendency to subsume livestock economics within 'agricultural economics' that continues till date. Hence once in a while one may come across some academic paper in *Indian Journal of Agricultural Economics* that touch upon aspects of livestock economy of India. This tendency to look at cattle as second to crops—livestock as second to agriculture—echoes how the veterinary departments and officers were treated during the late colonial period in British India. While a Board of Agriculture came into being in 1905, proposals to constitute a similar Board to be a policy making body on animal husbandry concerns were not attended to, even as they were voiced by veterinary officers. The division of labour debates amongst the officers of Department of Agriculture and Civil Veterinary Department were intense and one can witness 'emotional investment' that informed the tenor of those debates.
9. See for more details, www.rln.org.in/our-approach/.
10. A few representative examples of such writings are: Swaminathan, Madhura & Yoshifumi Usami (2016) 'Women's Role in Livestock Economy', *Review of Agrarian Studies*, Vol 6, No 2, pp. 123–134. https://ras.org.in/697c22e48e9cb0207d3e5d7560c971f6.
11. See, Prasad, Archana (2006) 'Review of *Ecology, Colonialism and Cattle*', in *Social Scientist*, Vol 34, Nos 11–12, pp. 80–82.
12. Baber, Zaheer (2008) 'Review of *Ecology, Colonialism and Cattle*', in *Agricultural History*, Vol 82, No 1, pp. 109–110.
13. See for detailed argument, Adcock, Cassie & Radhika Govindrajan (2019) 'Bovine Politics in South Asia: Rethinking Religion, Law and Ethics', *South Asia: Journal of South Asian Studies*, Vol 42, No. 6, pp. 1095–1107.
14. See *Proceedings of the Second Cattle Conference* (1937), Government of India Press, Shimla, p. 04. Similar views that stressed upon the centrality of ox in agricultural landscapes and lives in India gets expressed by W. Smith, Imperial Expert of Dairying multiple times in his writings. For example, speaking at the second meeting of Veterinary officers in India, he states, "The question of cattle was the most universal one in India and no matter what crop particularly

interested the agriculturist, he was everywhere dependent upon the efficiency of the ox as the main factor in cultivation", *Proceedings of the Second Meeting of Veterinary Officers in India* (1923), p. 106.

15. Mishra, Saurabh (2015) *Beastly Encounters of the Raj: Livelihoods, Livestock and Veterinary Health in North India*, Manchester University Press, Manchester, p. 02.
16. Elsewhere we find historians of medicine engaging with epizootics in early 1970s. See, Onselen, C. Van (1972) 'Reactions to Rinderpest in South Africa, 1896–97', *The Journal of African History*, Vol 13, No 3, pp. 473–488.
17. See, Curth, Louise Hill (2002) 'The Case of the Brute Beast: Animals and the Seventeenth Century Animal Market Place', *Social History of Medicine*, Vol 15, No 3, pp. 375–392. Also see, Mishra, Saurabh (2014) 'Introduction: Veterinary History Comes of Age', in a virtual issue of *Social History of Medicine* on the theme of Veterinary History (September 2014), http://www.oxfordjournals.org/our_journals/sochis/veterinaryhistoryintro.pdf.
18. Casimir, Machael and Aparna Rao (2004) *Nomadism in South Asia*, Oxford University Press, New Delhi.
19. Howard, Albert and Gabrielle L C Howard (1929) *The Application of Science to crop Production*, Oxford University Press, Bombay.
20. Howard, Albert and Gabrielle L C Howard (1929) *The Development of Indian Agriculture*, Oxford University Press, Bombay.
21. While there is a need to an intense engagement on the Indore experiment undertaken by Sir Albert and Gabrielle Howard along with their colleagues, it is a pity that there is no intellectual biography on Sir Albert Howard that can give us an insight into how his Indore experiment acted as a crucible for his early ideas on Organic Agriculture. In 1978, when Ian Carruthers made an ardent plea to undertake such an academic research, it was observed that the very institute whether Howards worked was not traceable except for the fact that one can acknowledge that here lies a place where Howards lived in 1920s. Carruthers concludes that note by asking, "Can India afford to sustain scientists in narrow specialism, to avoid village contact, to neglect biological husbandry and the insights to be gained from studies in archive?" See, Carruthers, Ian (1978) 'Raking up Mud and Magic for Agricultural Progress', *Economic and Political Weekly*, Vol 13, No 20, p. 834.
22. Keen, B. A. (1932) 'The Real Problem in India', Paper read at the Royal Institute of International Affairs on 1st March 1932. Quoted in Wright, Norman C. (1937) *Report on the Development of Cattle and Dairy Industries of India*, Government of India Press, Shimla. p. 50.
23. See, Pereira, Charles (1982) Bernard August Keen, (5 September 1890–5 August 1981), *Biographical Memoirs of the Fellows of Royal Society*, Vol 28, pp. 205–223.
24. See, Bryce Burt's opening and introductory remarks to Arthur Olver's Lecture on 'Animal Husbandry in India' at Royal Society of Arts in 1942 (*Journal of Royal Society of Arts*, Vol 90, no 4614, pp. 433–451. Also see, Burt, Bryce (1942) 'Agricultural Progress in India during the decade of 1929–1939', *Journal of Royal Society of Arts*, Vol 90, No 4607, pp. 202–223. For an extensive discussion on mixed farming experiments, see *Proceedings of the Second Meeting of the Crops and Soil Wing of the Board of Agriculture and Animal Husbandry in India* (held at Lahore from 6 to 9th December 1937), Government of India Press, Shimla.
25. See, 'Nutrition in relation to Human Beings, Farm Livestock and Crops', *Agriculture and Livestock in India*, Vol VI, Part 3, (May 1937) pp. 388–393.

Chapter 2
Revisiting 'Surplus Cattle' and 'Sacred Cattle' Discourse: Why a History of Cattle in India Needs to Move Beyond This Framing?

As argued in the introduction, academic writings on cattle by economists have remained occupied with the notion of 'surplus cattle' for many decades now. During early years of the twentieth century, papers written about cattle development in India repeatedly used two phrases, 'surplus cattle' and 'sacred cow', so much so that a large part of livestock economy literature even in post-independence India, including economic and anthropological works, continued to grapple with the 'problem of numbers in cattle development'.[1] In his survey of the concept, a social scientist quotes an impressionistic account noted by John Voelcker, whom Government of British India had invited to suggest improvements in Indian agriculture around late nineteenth century.[2] Voelcker was impressed by a large number of Brahmini bulls wandering about and had described them as 'standing religious menace' to crops. His impression that the number of such breeding bulls might be in excess to numbers needed to fulfil breeding needs of cattle in India had found validation when the first livestock census in British India in 1919–20 got published and reported that there were 5.1 million breeding bulls in a total cattle population of 113 million.[3]

Curiously, even as the surplus cattle discourse emerged in the wake of that remark by Voelcker, at the turn of the century, owing to frequent famines in the last decade of nineteenth century, there had arisen complaints that frequently drew attention to diminution in the number of plough cattle. Native peasants and their representatives cited many reasons for such a perception, such as breaking-up of grazing grounds, the increase of forest dues, the spread of epizootic disease, and the indiscriminate and excessive slaughter of cattle.[4] However, a thorough understanding of the existing situation was constrained by the fact that statistical information regarding agriculture and livestock over a large proportion of British India had attained any real approximation to accuracy only from 1919 onwards when Livestock census started to be carried out throughout British India and native states. In the year 1912, the imperial government had not yet launched an all-India Livestock Census and much of the agricultural statistics and cattle number available till then came from returns filed with the Land and Revenue Departments. Still there existed a feeling amongst

H. Upadhyaya, *Critical Insights on Colonial Modes of Seeing Cattle in India (1850–1980)*, Asia in Transition 27, https://doi.org/10.1007/978-981-97-1560-2_2

colonial experts that for some provinces such as Bengal the figures on cattle were highly unreliable, while the figure for Madras and Bombay were believed to be more accurate.[5]

The agricultural statistics that became available in the year 1912 had given an imperial expert on agriculture, C. E. Low a general impression that the numbers of cattle have more than kept the pace, over India as a whole, with the cropped area. However, he was cautious not to take refuge in statistics. Low underscores the crucial crop-cattle interdependence that was obtained during the first decade of last century, when he states:

> It is clear that there is a limit to the number of acres which a pair of oxen can cultivate. If the number of cattle fell short, by any appreciable quantity, of the number required to till the area under crop, it is clear that the number of acres tilled per head of cattle cannot be indefinitely increased, and there must inevitably be, sooner or later a decline in cropped area, unless the agricultural practice materially deteriorates, or the quality of the cattle greatly improves, neither of which, it is well known has been the case. On the contrary, the cropped area, in spite of the occasional setbacks due to famine, has been steadily rising, and this shows plainly enough, that there has been no decline in the number of cattle.

While it is clear from the above statement that Low was reporting the situation vis a vis draught cattle and doesn't engage with the question whether the number of cows could be considered as surplus, contemporary accounts from Punjab do hint at the strains arising from the rapid expansion of cultivation that had resulted in the shrinkage of grazing and had started to impact the tracts where earlier local breeds were bred.[6] In several writings, this line of argument 'cropped area per pair of plough bullocks' was put forward to hint at the deficit of draught cattle. By comparing the numbers of cattle relative to population and area, Nilananda Chatterjee (1926) sought to argue that relative to the area and population served the number of cattle that India supported was not very high. Next, he made an argument that if one supposes that a pair of bullock is needed to plough 5 acres, then both British India and Native States showed a deficit of plough cattle rather than a surplus. Similarly, he went on to show that the ratio of milch cattle to persons also left much to be desired. Tables 2.1, 2.2 and 2.3, Chatterjee presents were advanced in order to push a 'plough cattle are in deficit, not in surplus' and 'more milk, not more beef' line that was adopted by All India Cow Conference Association founded in September 1917 and patronized by a British Orientalist, Sir John Woodroffe (1865–1936). This organization, All Indian Cow Conference Association was founded "with the aim of addressing the problems of India's cattle resources from a strictly economic point of view" (Adcock 2019: 1145).[7] While not much has been written about what prompted the then British justice of Calcutta High Court to patronize the cause of cow protection, Adcock (2019: 1145) hints from the little material that is available that "the first, preliminary meeting of the AICCA was held in Calcutta on 29th September 1917". Drawing from the fleeting reference to this aspect of John Woodroffe's personality in the biography authored by Kathleen Taylor, Adcock notes that "Taylor's biography notes Woodroffe's involvement in the AICCA but offers few details about the organization". Did Woodroffe get involved in espousing and patronizing the cause of Cow Protection, due to his deep interest in Tantra philosophy and his intellectual works on

Table 2.1 Proportion of plough cattle to cultivated area in the year 1926–'27

Region and year	Cultivated area in acres	Total no. of plough cattle	No. of acre actually available for cultivation	Quantity of cultivated area in acres per plough cattle pair	Remarks
British India 1914–'15	227,611,000	48,645,000	24,322,500	10	One pair of bullocks in India can till only 5 acres per working season
Native States 1914–'15	31,935,000	4,002,000	2,001,000	16	
Total and average	**259,546,000**	**52,647,000**	**26,323,500**	**10**	

Source Chatterjee, Nilananda (1926) 'A Note on the Cattle of India' in *The Condition of Cattle in India*, All India Cow Conference Association, Calcutta

Table 2.2 Cropped area per pair of plough cattle in principal states in Kathiyawar

Year	State	Number of Bullocks	Total cultivated area (in acres)	Cropped area per pair of Bullock (acres)
1934–'35	Bhavnagar	55,529	735,538	26.5
1933–'34	Jamnagar	82,636	1,098,286	26.6
1933–'34	Junagadh	98,767	682,413	13.8
1933–'34	Rajkot	14,213	107,618	15.0
1933–'34	Gondal	29,064	352,006	24.2
1933–'34	Porbandar	28,335	163,862	11.6

Source Trivedi, A. B. (1943) *Kathiyawar Economics*, A. B. Trivedi, Khalsa College, Bombay p. 104

Tantra, which were published under the pseudonym, Arthur Avlon? Did he agree to preside over AICCA since those intellectual voices behind this strand of Cow Protection movement committed itself to articulate the issue "from a strictly *economic* point of view" as opposed to *religious* viewpoints against the cattle slaughter that were articulated earlier in the late nineteenth century by Arya Samaj leader Dayananda Saraswati? AICCA had Justice W. E. Greaves as its second president and a 1926 publication carries the photographs of both Woodroffe and Greaves.

In 1926, writing an introductory essay to Nilananda Chatterjee's book titled, *The Condition of Cattle in India: Being an Inquiry into the Causes of the Present Deterioration of Cattle with Suggestions for their Remedy*, Woodroffe recounts how he was approached by Mr B. Chakrabarty (who practised law at Calcutta High Court) and other likeminded gentlemen who presented to him the idea of "founding an association with the object of protecting and improving cattle on purely economic basis" and how he "readily consented to be its President".[8] In this introduction, he also stressed upon that it was his suggestion to launch a comprehensive inquiry by "sending a circular letter to all District Boards, Municipalities, Agricultural Associations, Gowshalas and other interested bodies" that requested these recipients to

Table 2.3 Cropped area per pair of plough cattle for some of the districts in Gujarat

District	Cropped area per pair of Bullock (acres)
Ahmedabad	24.8
Kaira	16.2
Panchmahal	10.5
Broach	22.8
Surat	11.8

Source Trivedi, A. B. (1943) *Kathiyawar Economics*, A. B. Trivedi, Khalsa College, Bombay p. 105

furnish replies to "a set of questions enquiring about the condition of local cattle and milk supply" as well as to provide their "suggestions for (cattle) improvement". He goes on to explain that after receiving several replies and concerted efforts at follow-up for two years to obtain a fairly representative sample, AICCA had formed "committee with Col. A. Smith as Chairman" and assigned them the task of considering the replies received. AICCA then "thought it advisable to present the public with a summary of the replies received" in the form of a well-curated and commented publication and assigned the task to Nilananda Chaterjee. Pointing out AICCA's connection with Bengal Humanitarian Association, Woodroffe stated in his introductory essay that "(Chatterjee's) connection with the Bengal Humanitarian Association for several years has helped him to carry out successfully researches in cattle breeding and cattle feeding, and he has thus gained valuable practical experience which he has not grudged to place at the service of the reader".

This publication by AICCA was followed by another "monograph, whose primary purpose was to brush away popular misunderstandings regarding the many side-issues of the problem of Cow Protection in India" from The South Indian Humanitarian League, the next year, in 1927. It was penned by L. L. Sundar Ram, a fellow of the Royal Economic Society, London and presented a comprehensive account of attitudes towards the cause of cow protection found in different religious texts and talked about the arrangements and official decrees that prohibited the cattle slaughter in past under different rulers before the arrival of British in India.[9] One of the chapters from this book, titled 'The Indian Attitude towards Cow Protection' was published in October 1926 issue of the monthly magazine *Mahabodhi* founded by the Angarika H. Dharmapala in 1892.[10] This indicates that L. L. Sundar Ram's effort at historicizing cow protection movement in India had found some traction from the Mahabodhi Society and amongst the movement that aimed to revive Buddhism in India.[11] It seems Sundara Ram (also known as Lanka Sundaram) was planning to publish the second volume of his monograph on Cow Protection in India, dealing with the economic side of the problem.[12]

Similarly, the Humanitarian League of Bombay published a similar tract titled, *The Cattle Problem of India* authored by Maganlal M. Shah in 1927. By early 1930s, the Humanitarian League of Bombay was actively engaged in the propaganda by instituting an essay writing competition for graduate students of Bombay Presidency on the theme of 'cow protection'.[13]

Similarly, giving the statistics on the native states in Kathiyawar region for mid-1930s, A. B. Trivedi argues out that "the insufficiency of the bullocks as well as implements is obvious, when a farmer tries to work out an area of 20–25 acres with a pair of bullocks". The statistics that he reproduced in his book for princely states in Kathiyawar are shown in Table 2.2 and having presented this numbers, he goes on to compare this with the situation that existed in some of the districts of Gujarat:

"Not only was the question in Kathiyawar confined to the realm of statistical deficit", Trivedi states, "coupled with the inadequacy of the plough cattle, is the growing complaint of its inefficiency".

In contrast to these arguments from native elites, in the imagination of colonial experts there was a strong view that native peasants and nomads practised 'overstocking' owing to the religious believes that considered cow and her progeny to be sacred. The employment of the symbolic value of 'sacred cow' as a tool—some political commentators have used the phrase 'pawn on a political chess board'[14]—and the surge of 'cow protection' discourse provided fuel to the Malthusian simplifications that informed the 'surplus cattle' discourse. Perspectives that had roots into middle-class anxieties about the adulteration of milk and milk products accompanied by soaring prices had by now started to put a heavy premium on perceived notions about the poor milk yields of zebu cows, rather than trying to engage with the resilience, heat tolerance and endurance as qualities in draught bullocks in India. Advocates of 'surplus cattle' discourse also failed to pay attention to the great odds against which the nomadic cattle breeders were trying to carry on with the cattle breeding to meet the increasing demands from rapidly expanding irrigated and cultivated tracts when both frequent famines and cattle plagues took a heavy toll of their bovine stock. As a result, there emerged concepts such as 'overstocking' and 'overgrazing' rooted into the prevailing ideology of Malthusianism that had very little space for self-doubt and nuances. Low (1912: 331–342) clearly appears to be articulating such a perception that heavily criticized the tendency of native peasants for overstocking, when he writes:

> Those officers who have had experience of severe famine conditions know well how immediately a fall in the number of cattle is followed by a fall in cropped area. The question is not entirely one of statistics, however. For the total number of cattle reported includes, and always has included, a large number of useless beasts that eat up the food required by the more valuable animals. Any cause that tends to diminish the number of such animals is a positive benefit to Indian agriculture, and though it may be a matter for regret that the slaughter of even a small proportion of cattle that meet this fate every year should cause pain to our Hindu fellow-subjects, it is far from being injurious to Indian agriculture.[15]

However, by the time when Low was expressing this view, the discourse around 'cow protection' and the criticism of the ill effects of slaughter had already started to gather wind. Adcock (2019: 1143–44) alerts us to relatively less attention that has been paid in academic scholarship on Cow Protection in India about "the alliance of the state with cow protectionist organizations during the colonial period" and traces how the cow protection discourse had started to receive support from "the highest ranking British officials in the Government of India" in early decades of twentieth century and how "colonial state's alliance with cow protection was forged in the name of breed improvement".

The argument 'sacred and hence surplus' was now being confronted by the argument put forward with great emotional investment by cow protectionists who wielded the figures on the expansion of trade in hides and dried meat. A British judge from Calcutta, Sir John Woodroffe, had asked, "if the twentieth century deterioration of cattle in India resulted from excessive rather than insufficient slaughter".[16] Cow protectionists were now increasingly relying on making arguments that revolved, not around the sentimental views of the place cow held in Hindu belief system, but on the view that talked about the impact of "excessive cattle slaughter" on material interests of Indian peasants. Trying to draw our attention to this discursive turn within the Cow Protection Movement, Adcock (2019: 1143) states her research shows a direction different from the earlier academic researchers, who had believed that the involvement of Cow Protection Movement into modern activities such a breed improvement was post-independent development. Adcock (2019: 1143) argues:

> Breed improvement has been integral to the political of cow protection since the early decades of the twentieth century. Integral, because it has formed a central component of cow protectionist argument and activity. But, above all, integral, because breed improvement made it possible for cow protectionists to secure the imprimatur of the state to their efforts, apparently vindicating cow protection's central claim: that their concern with cattle is not a matter of private Hindu sentiment but is in the material interest of the country as a whole.

Drawing our attention to what she calls "neglected body of sources", Adcock (2019: 1144) argues against the fallacious distinction often presented by colonial officials "between the Hindu religious attitude towards 'sacred cow' and British 'scientific' attention to cattle as livestock". Presenting her review of these hitherto neglected sources, Adcock (2019: 1144) asserts, "Not only did cow protectionists speak the language of colonial science, but the colonial state embraced cow protectionist institutions".

Around the time when Chatterjee's book got published by AICCA, in addition to the sentiment that the expansion of trade in hides and dried meat was driving the excessive slaughter of cattle, there arose another sentiment that blamed the city-based milch stables as the cause for 'genetic drain of high yielding milch breeds of cattle'. Organizations such as the Humanitarian League, Bombay actively pursued this line of argument and tried to influence upon the policy makers to put in effect restrictions and regulations on slaughter, even going to an extent of demanding a prohibitory law.[17]

Another British veterinary expert, J. Matson had found the Indian cattle so helplessly inefficient that he considered them as having no commercial value beyond that of their hides, and thus he argued that these cattle represented a liability, not an asset; a burden, not a wealth.[18] However, such an opinion by Matson came with a qualification. In fact, the very next sentence that eager articulators of 'surplus cattle' discourse have often conveniently avoided paying attention to. Qualifying his very strong views about "hopelessly inefficient cattle", Matson had stated, "The fact is commonly ascribed to overpopulation and that is the chief cause, but recent observations suggest that *it is by no means the whole story; that in fact, if there were no disease, a substantial proportion of the useless emaciated specimens one sees would be marketable cattle*" (emphasis added). This qualification by Matson reporting from Central Provinces forests was also accompanied by a fervent plea for providing veterinary services to India's livestock rearing peasants and nomads and wished that British Government committed more funds for the same. However, dozens of academic papers cite from the section where he refers to the inefficient and useless cattle, without realizing that his remarks were not applicable to all cattle belonging to all different breeds everywhere in India and his remarks did follow a significant qualifier exposing the colonial government's failure to spend substantially on veterinary health of an ordinary Indian peasant's cattle. Even as the academic literature on 'surplus cattle' discourse has been prolific, it does also betray a tendency to exhibit its own blind-spots, such as the one identified above.

A corollary view to the 'surplus cattle' discourse that often finds expression in literature on livestock economics in India is that drastic changes in cattle management were needed and India must get rid of the burden of those inefficient cattle by the way of slaughter rather than the slow process of seemingly natural death by what has been referred to by some commentators as 'deliberate starvation'. However, despite the refrain on 'surplus cattle', colonial policy had not always advocated indiscriminate cattle slaughter at all the times. As Adcock (2019: 1149) points out,

> The 'waste' of high-quality breeds in the cities was acknowledged by the colonial government to be a problem that impinged on government schemes for breed improvement since it caused depletion of the resources needed for scientific breeding schemes and undermined efforts to upgrade country cattle… Colonial officials did come to endorse cow protectionist assistance in the 'salvage' of dry cattle from the cities, or what came to be known simply as 'urban salvage'.

Adcock (2019: 1150) cites a statement published in the *Agricultural Journal of India* in 1922, where in "the imperial dairy expert, W. Smith, had this advice for advocates of cow protection: 'the first and most needful form of cow protection urgently wanted in India is the stoppage of the slaughter of young cows and female buffaloes in the cities'".[19]

There had been a few occasions when colonial administrators had taken an anti-slaughter view, even if rare. There are some accounts that tell us about the enormity of cattle slaughter during the food crisis around World War II and the decision by colonial government to ban cattle slaughter. An account presents us details of how increased consumption of beef by the armed forces accompanied by famine conditions in Bengal led to higher prices for meat and food grains, when rice imports to British India were cut off due to Japanese occupation of Burma.[20] When the slaughter of cattle started to prove detrimental to crop-livestock feedback system by a near total destruction of milch animals, draught animals and bullock replacing animals, a direct military intervention was sought. In the summer of 1944, colonial government banned cattle slaughter within the jurisdiction of the Bengal, as explained by the following statement from Knight (1954: 141):

> During the war there was an urgent need to reduce or to avoid the slaughter for food of animals useful for breeding or for agricultural work. For the summer of 1944, the slaughter

> was prohibited of: 1) Cattle below three years of age, 2) Male cattle between two and ten years of age which were being used or were likely to be used as working cattle, 3) All cows between three and ten years of age, other than cows which were unsuitable for breeding offsprings, 4) All cows which were pregnant or in milk.[21]

The account presented in the above quote by Knight gets reaffirmed by another researcher Lensch (1987: 55) too:

> In 1944, West Bengal had to witness acute starvation on account of a prolonged drought period, on the one hand, and Japanese occupation of Burma, on the other, cattle slaughter had assumed such proportion that British troops had to be called in to enforce the implementation of the ban on slaughter. Without this intervention, the West Bengal agriculture, which played a significant role in catering for the needs to the allied troops that were engaged in war efforts for re-conquering Burma, would have collapsed.[22]

Similar accounts are presented in other contemporary sources on how the slaughter of Gujarat's milch stock was resorted to feed the influx of the alien soldiers during the Second World War and restraining orders had to be issued by authorities immediately thereafter.[23]

2.1 Efforts to Estimate 'Surplus Cattle'

During the second phase of surplus cattle debate between 1920 and 1950, the arguments were followed by estimations based on available statistics and experimental data. Economists C. N. Vakil and S. K. Muranjan wrote in 1927 that slaughter of one-fifth of existing cattle would not have any detrimental effect on food grain availability.

While giving his evidence before the Indian Taxation Enquiry Committee (1925: 359), Sam Higginbotham, a Presbyterian Missionary pronounced his opinion that "the worst economic loss is due to the surplus cattle in India". Referring to the Cattle Conference held a year ago (i. e. in 1925) in Bangalore, he reported that "the surplus cattle costs India 4 to 9 times the land revenue".[24] In 1938, Radhakamal Mukherjee (1938: 132) had remarked that "the number of livestock (meaning cattle) could be safely reduced in the whole of India to one-third of the present population, without affecting the standard of farming and rural transport". In his book, *Food Planning for Four Hundred Million*, Mukherjee had regarded 125 million heads of cattle as "uneconomical and superfluous".[25]

Writing in 1941, an essay titled 'Animal Husbandry and Crop Planning in India', K. C. Sen and S. C. Ray acknowledged that India continued to face the problem of "large number of uneconomic cattle". Sen and Ray worked with the Animal Nutrition Section of Imperial Veterinary Research Institute, Izatnagar and in this essay they made a fervent plea to recast the existing agricultural policy. In this article, Sen and Ray stressed upon the need reorient the crop planning with a view to ensuring that the animal nutrition issue can be addressed by stating, "The ultimate objective of crop planning is the improvement of national health and prosperity. In a country such as India, successful crop planning is intimately associated with a healthy livestock industry". In concluding paragraph of this article, Sen and Ray (1941: 689) argued:

> It must be understood that in this country, animals and human beings are constantly competing for the available land and with the tendency towards an increasing population, this competition will grow further. It is true that we have a large number of uneconomic cattle. The disposal of surplus cattle in India is a problem which, while demanding urgent attention, is beset with many difficulties. But, sooner or later, we shall have to face the problem and a balance has to be struck somewhere to enable us to obtain a population figure commensurable with the food supply.
>
> It is probable that the ignorance on the part of the cultivator has led to the large increase in this uneconomic cattle population, which cannot be kept in health and productivity with existing food supply. The situation has been aggravated by the lack of a suitable agricultural policy. So far as the farmer is concerned, his ignorance is perhaps excusable but it is the duty of those, who are responsible for India's agricultural policy, to see that the crop planning of the country not only serves the needs of the human population but also that of the enormous livestock population on which the continued prosperity of the agriculturist so largely depends.[26]

An expert team sponsored by Ford Foundation, which was assigned the task to recommend how to tackle food crisis in India in late 1950s compared bovine population to feed availability to arrive at a similar conclusion. The Ford Foundation team (1959: 64) talked about "widespread recognition not only amongst animal husbandry officials, but among citizens generally, that India's cattle population is far in excess of the available supply in fodder and feed". It reached at conclusion that, "At least one-third and possibly as many as one-half of the Indian cattle population may be regarded as surplus in relation to feed supply".[27]

Just a year before this, R. O. Whyte (September 1958: 203), an FAO expert and the then Fodder and Grassland Advisor to Government of India had written:

> Now in modern India, the animal management is very poor and semi starvation of most of the cattle population for most of the year is the rule. Milk production per animal is steadily decreasing, the intervals between calving are excessively long and the cows have difficulty in feeding their calves. The great and overriding problem is the presence of the large number of the so called "surplus animals", aged animals, bullocks too puny for work, and cows which are largely infertile, or of little value for rearing purposes. The absence of cow slaughter and uncontrolled breeding merely tend to increase the numbers and so also competition for the very limited feed supply available.[28]

The following decade of 1960s witnessed several economists estimating the number of surplus bullocks and cows. I. Chatterjee (1962: 773) tried to estimate surplus and suggested that reducing the number of adult female cattle (cows) by about 20 percent of their prevailing level would result in an adult female cattle (cows) population of about 60 percent of the adult male cattle.[29] Chatterjee's estimation appears to be very conservative compared to that of Dandekar's (1964: 353), Whyte's (1967: 168), Heston's (1971: 193) and Mishra's (1978).[30] Whyte (1967: 168) estimates the number of surplus cattle by basing his perspective squarely on the role of cows in milk production stating, "For economic milk production in India, only 25 percent of the present population of female zebu (cows) of breedable age is worth maintaining and feeding". Whyte's opinion was clearly influenced by predicting a larger role for buffaloes in maximizing milk production in India and he stated that almost 80 percent of buffaloes in India are worth maintaining.[31]

2.2 Cattle as an Integral Part of Agricultural System

Thus, we find a range of statements that label cattle population in India as surplus and some of these estimates suggest that the religious beliefs that hold cows as sacred and proscribe slaughter of useless cattle impede the efforts at cattle breeding with an aim for increased and 'economic' milk production in India. However, if one tries to search for the scientific literature that had looked at the milk production potential of the indigenous cows in village herds, one finds probably very, "little concrete information available on the milk production of Indian cows, outside a relatively few animals in government-owned or sponsored herds", as noted by livestock development expert from United States Department of Agriculture, R. W. Phillips.

A very early analytical report penned by Imperial Animal Husbandry expert, Arthur Olver and M. Vaidyanathan, estimates the total annual production of milk in India at over 1000 million maunds during the early years of 1930s.[32] Wright compares this citation with results of provincial milk and milk products marketing surveys that indicated the figure of annual production of milk must have been close to 700 million maunds, instead. In terms of monetary value, annual production of milk and milk products amounted to Rs. 300 crores at the then prevailing price levels, being roughly equal to the value of India's total output of rice and was three to four times the value of output of wheat.[33] Phillips also talks about just one exception, a survey by Imperial Council of Agriculture Research (1939) of village cattle in seven typical breeding tracts during the winter months.[34] Presented below are the major findings of that ICAR survey of seven prominent breeding tracts:

> The average daily milk yield observed for cows was 3.74 lb, and the estimated yield per lactation was 943 lb. Water buffaloes were also included in that survey and the figures for daily milk yield and per lactation were 7.82 lb and 2160 lb respectively. The average milk production per capita in the areas studied was 23.36 oz, while the consumption of milk per capita was 10.16 oz. This included *Ghee*, Curd and other milk products expressed as milk equivalent. The remaining 13 oz per capita got sold.[35]

Given that the data on milk production and per capita consumption from the above ICAR study came from the seven tracts where cattle breeding was an important pursuit, Phillips (1945: 403) speculated that "in other areas where breeding is of less importance and many working bullocks are purchased, the available milk supply would undoubtedly be lower". He then goes on to quote an authority (without naming the person) who had estimated that the supply of fluid milk for the whole population is less than one and one-half oz per day.[36]

Wright (1937: 155) points out that per capita production and consumption figures of milk and milk products range between 7 to 8 oz per day, assuming both liquid milk and milk products expressed in milk equivalents to be 800 million maunds (i.e. 6400 million gallons). Wright does concede that since these figures are "based on very rough estimates of the average milk yields and lactation periods of Indian cattle," one may ask to what extent it represents true indication of per capita production and consumption. He also indicates that "an alternative method of determining the level of milk consumption" in India was to rely on the statistics available from dietary surveys. However, he notes that "unfortunately few such surveys have been carried out". On the question of milk yields of village cattle, he had relied on the 'Results of Provincial Marketing Surveys' carried out by Dara N. Khurody and reached a conclusion that the average typical yield in the case of British India was "only a little over 600 lb per year, which would represent an average daily yield of say, 4–6 lb." In Provincial Marketing Surveys, Punjab had recorded the highest average yields per lactation, viz. 1400 lb. for cows and 2160 lb for buffaloes and Assam the lowest, viz. 170 lb for cows and 430 lb for buffaloes.

K. R. P. Kartha, a well-known expert for his work on lactation records at government dairy farms in his presentation at the meeting of Animal Husbandry Wing of Board of Agriculture (February 1933) presented that average lactation yields were higher in case of institutional herds (those maintained by government breeding farms and military dairy farms), wherein the lowest average lactation yield recorded by ordinary Sahiwal breed was 3864 lb. and the highest average lactation yield was recorded by Half-bred cows at 6500 lb. The average lactation yields for buffaloes, Sindhi breed of cows and Ferozepur Sahiwal were 4000, 4150 and 5717 lb, respectively. This data does point out that a few indigenous breeds had clearly emerged as better milk yielders by 1930s.[37]

So, if that was the case with statistical data on milk yields of cattle and buffaloes in India in 1930s, one can guess how much statistical data existed that would aid an understanding on contribution of cattle to 'agrarian economy' by providing services such as draught and carting as well as manure! The discourse on 'surplus cattle' during the decade of 1960s—when right-wing political parties had heightened the 'cow protection' agenda demanding a nationwide ban on cow slaughter—frequently put forward the Hindu religious view that regard cow as sacred to be responsible for giving rise to the problem of numbers. However, during the decade of 1920s, the statistical data on the export of hides, skins and offal showed India as being the largest exporter of hides and skins in the British Empire. Duckham (1932: 222) showed that in 1928, the total exports of hides and skins from colonies were valued at £ 42,200,000 of which India's share was £ 14,000,00 (i.e. to say as high as 33.17 percent), roughly Rs. 40 crore at the then prevailing prices and currency conversion rates.[38] Other contemporary sources also point to a massive boom in the leather export market and put the value to have risen forty times between 1860 and 1925, leading to increase in cases of cattle poisoning, with cases being reported from the Madras and Bombay Presidencies also.[39] Despite the religious taboo on beef eating and religious belief that spurred agitations against the cattle slaughter, it was British India that was the major driver of the leather industry of Empire! While definite statistics on the number of slaughterhouses in existence in British India and the amount of slaughter is not available, a recent publication by a Gandhian scholar Dharampal (2002: 451) cites from a report of the Madras Presidency published in 1893 making a claim that there was a beef slaughterhouse in each district in the Presidency.[40]

Dara N. Khurody (1974: 11) quotes the statistics available from the year 1967–68 on the contribution by livestock sector to national income and provides in Table 2.4.

The evaluation of the importance of cattle in Indian agriculture would be incomplete unless an effort is made to understand their contribution to on-farm agricultural operations, going beyond looking merely at the valuation of 'animal products'. Wright conceded that while it was found to be quite "difficult to place a definite monetary value on cattle labour," rudimentary efforts were made to assess "the value of cattle labour in holdings in the Punjab," under the direction of Imperial Council of Agricultural Research. The research in Punjab had showed that "between 15 and 20% of the cost of cultivation" could be attributed to cattle labour. He extrapolated this to all India level to arrive at the share of cattle labour to be somewhere between Rs. 300 and 400 crores, at the then prevailing price levels. However, this was an extrapolation; Wright also used an alternative methodology from the Royal Commission on Agriculture report. The Royal Commission on Agriculture (1928) had estimated the average cost of feeding and depreciation of a pair of bullocks at Rs. 175 per year and made an assumption that a pair of bullocks is needed to cultivate 10 acres of net sown area.[41] Assuming 'net sown area' at 230 million acres in British India and 70 million acres in Indian states, he arrived at the cost of cattle labour to be Rs. 525 crores for all India and Rs. 400 crores for British India. He also cites that early study by Olver and Vaidyanathan, while discussing "another indirect source of income attributable to cattle, which is difficult to assess," namely their contribution to the fertility of the soil. Olver and Vaidyanathan had attempted to arrive at a rough estimate of the cash value of cattle manure suggesting a provisional figure of Rs. 270 crores at the then prevailing prices. He also voiced his concern that "No figures exist to show what proportion of cattle dung is used as fuel, although this is believed to be very large".[42] During early 1960s, those writing about agrarian history of India stated that "the existence of surplus or useless cattle cannot be determined separately from the agricultural system in India, *of which they form an integral part, providing most of the draught power, fuel and food to farming population and contributing about 50 percent of the total income from agriculture*" (Randhawa 1962: 259, emphasis added). Randhawa's opinion was not very different from a stated official position of agricultural research establishment in India in 1962. ICAR (1962: 09) had clearly stressed in the same year that "if milk production is unduly pushed up, it may cause deterioration in the draught quality of the cattle and indirectly affect the entire basis of agriculture in India".

Table 2.4 Estimation of the value of output from cattle

Milk production function	22 Million tonnes @ Rs. 800/tonne	Rs. 17,600 Million
Cattle labour	40 million pairs @ Rs. 4/day X 150	Rs. 24,000 Million
Dung etc.	800 Million tonnes @ Rs. 35/tonne	Rs. 28,000 Million
Beef, hides, bones, etc.		Rs. 500 Million
Total		Rs. 70,100 Million

Source: Khurody, D. N. (1974) *Dairying in India*

Hanumatha Rao (1988: A 142) considers the phase of Indian agriculture up to the mid-1960s as the extension of cultivation as strategy for growth rather than increasing yield per unit, i.e. 'intensification'. Accompanied by "the failure of land reform measures to effect a significant redistribution of land to the landless and the political will of a democracy to provide wherewithal for the landless," this strategy had contributed substantially to the encroachment on village common lands and forest lands by bringing them under the plough as outlined by Nadkarni (1987). T N Khoshoo (1986) in his presidential address to the annual meeting of Indian Science Congress Association estimated that out of the total forest cover lost in the country, as much as half the area has been brought under the plough. Hanumantha Rao (1988: A 142) links the extension of area under cultivation strategy with a significant increase in cattle population, since this strategy made demands on human labour and draught power in a region that had recorded sluggish growth in irrigation till mid-sixties. Groundwater-based irrigation in that phase was also largely delivered by bullock driven Persian Wheel.

Even as agriculture and cattle development in India were undergoing strategic shift by embracing intensification in mid-1960s, Marvin Harris (1966: 51) made a plea to study the question of 'surplus cattle' in India from the cultural ecology lens. Harris attempted to explain "taboos, customs and rituals associated with management of Indian cattle" by paying close attention to "positive-functioned and probably adaptive processes of the agrarian ecological system of which they are a part" rather than as a result of religious beliefs. Harris presented his hypothesis that "when viewed as part of an agrarian ecosystem rather than as a sector of a national price market, it is doubtful that any component of the cattle complex is 'useless'". Harris cites in his 1966 essay several opinions that underlined inadequacy of draught power available for land preparation at the busiest season of the year, even with alleged 'overstocking'. The arguments put forward by Harris triggered a long debate on 'the Sacred and Surplus Cattle in India' in pages of the journal 'Current Anthropology' and S. N. Mishra presents a survey of this lengthy debate in an overview article in 'Sociological Bulletin'.

2.3 Conclusion

This chapter has attempted to portray how the bovine stock in colonies such as India was perceived in expert writings by experts working within the science and technology institutions of late colonial state, as well as in writings by native experts. While trying to communicate the limitations of this framing, where writings on cows and cattle have often remained confined to the master narratives around the idea of 'Sacred Cow/Mother Cow' and 'Surplus Cattle', this chapter alerts us to the need to go beyond this binary in an effort to write the history of bovine animals in India from the perspective of how the lives of bovine animals were intimately tied with peasant lives in an agro-pastoral landscape.

With revisiting how the discourse on Surplus Cattle and Sacred Cow kept growing in multiple writings across time, this chapter highlights the concerns articulated by the High Modernist colonial state that undertook the endeavour of "counting the number of livestock" through institutionalizing the Livestock Census every five years. The chapter underlines divergent point of views on the "problem of numbers in cattle development in India" by situating the crop cultivation—cattle rearing relationality around the emergence of colonial science and scholarship that shaped discourses on Agriculture Science, Veterinary Science and organized dairying and teaching of Dairy Technology.

This chapter gives a hint on how meanings attached to cattle were set to be reimagined and reconfigured by invoking the binary of cash crops versus food crops/ fodder crops and similarly the binary of promising milch animals and plough animals versus the superfluous and non-productive (i.e. surplus) cattle, who encroached upon the availability of animal nutrition for the better specimen of the species, given religious attitude of Indian peasants, who made the agenda of cow protection a big political rallying point to other the religious minorities as well as colonial administrators based on their dietary habits of consuming beef (cow meat) and carrion. The meaning attached to bovine animals was also getting shaped increasingly by the science and technology institutions engaged in communicating the modern science, wherein the British Empire looked at Europe as the 'seed bank' and 'stud farm' of the colonies.

Alfred W. Crosby in his book, *Ecological Imperialism* has traced the biological expansion of Europe through transplantation of European diseases, flora and fauna in temperate zones of the world—North America, Australia and New Zealand. In the case of India and other colonies, however, the trajectory had been different from the one that Crosby's narrative presents about Dominions. As Deepak Kumar has shown in his writings on History of Science and Technology during the 1830–1900 period, in the eyes of colonizers, India presented an image of 'promised land' which was now being referred to as 'plantations' and 'native' peasants were being sought after as worthy subjects who could be made to appreciate the idea of 'settled cultivation' and the modern science of crop cultivation and cattle rearing for milk production. High Modernist State thought of earnestly undertaking scientific research on cash crops such as Cotton, Jute, Opium, Indigo and Tea, without battling an eyelid for a second on what would work for the native peasants.

This chapter has tried to argue how the history of bovine animals needs to go beyond the master narrative frame of 'Sacred Cow/Mother Cow' and 'Surplus Cattle' in order to understand and pay close attention to crop-cattle relationality having a manifest impact on shaping cattle keeping as well as pastoral and peasant livelihoods. It underscores how the notion of 'surplus cattle' so often invoked while speaking about the challenge in the agenda of 'cattle development' was not merely a product of statistical tyranny, but this very discursive frame was itself a construct that derived its origin from the conceptual semantic field operating within the halo of High Modernist State. While this chapter, revisits the moments of the emergence of 'surplus cattle' discourse running through multiple texts in writing by colonial experts as well as native elites gripped by the idea of 'modernizing' impulse, it is possible to trace the moments when colonial experts working within Science and Technology institutions of the High Modernist imperial state started expressing humility and eagerly offered their time to listen to and learn from native peasants.

Continuing this exploration to write history of bovine animals in India and understand what insights emerge on the colonial modes of seeing cattle, the next few chapters shall engage with how we are able to trace the ideological antecedents of 'green' and 'white' revolutions in our late colonial past (1850–1947). The next chapter engages with famines, cattle diseases (mainly Rinderpest) and cattle mortalities and how it all led to shaping of discussions on establishing civil veterinary departments in late colonial India.

Notes

1. Dandekar, V. M. (1964) 'Problems of Numbers in Cattle Development', *The Economic Weekly*, Annual Number, (February 1964), pp. 351–355. https://www.epw.in/system/files/pdf/1964_16/5-6-7/problem_of_numbers_in_cattle_development.pdf.
2. Mishra, S. N. (1973) 'Surplus Cattle in India: A Critical Study', Sociological Bulletin, No 22, pp. 297–308.
3. The first all India Livestock Census was held from December 1919 to April 1920, throughout British India, following the decision taken by Government of India in the year 1916. However, there have been references that suggest that counting of cattle had been a practice in certain regions right from 1850s. For the district Kheda (Kaira) the cattle figures are available right from 1840s. While the livestock census operations of earlier dates were confined to a few provinces, 1919–20 Livestock Census made 'counting of cattle' a well-coordinated exercise throughout British India and native states.
4. It is also quite curious to observe that even as several late colonial veterinary experts had referred to cattle population in India as 'surplus', historian Jairus Banaji makes a significant observation that "cattle were so integral to the peasant economy in nineteenth century that any household without it was a household on the verge of extinction'. See, Banaji, Jairus (1992) 'Capitalist Domination and the Small Peasantry: The Deccan Districts in the Late Nineteenth Century', in Gyan Prakash (ed), The World of the Rural Labour in Colonial India, Oxford University Press, Delhi, p. 124.
5. For the detailed discussion on the cattle numbers see, Low, C. E. (1912) 'The Supply of Agricultural Cattle in India', *The Agricultural Journal of India*, Vol VII, Part iv, pp. 331–342.
6. Department of Agriculture, Punjab (1910) *Cattle and Dairying in Punjab*, Department of Agriculture, Punjab, p. 05.
7. Adcock, Cassie (2019) ''Preserving and Improving the Breeds': Cow Protection's Animal Husbandry Connection', *South Asia: Journal of South Asian Studies*, Vol 42, No 6, pp. 1141–1155.
8. Woodroffe, John (1926) 'Introduction' to Nilananda Chatterjee's book, *The Condition of Cattle in India: Being an Inquiry into the Causes of the Present Deterioration of Cattle with Suggestions for their Remedy*, All India Cow Conference Association, Calcutta.
9. Sundara Ram, L. L. (1927) *Cow-Protection in India*, The South Indian Humanitarian League, Madras. The emotional investment of L. L. Sundar Ram in the cattle question in India can be understood, when we look at his criticism of the forest administration in India by the colonial government and his fervent plea to pay attention to Indian pastures and the question of fodder supply for cattle in India in his essay titled 'Indian Pastures and Fodder Supply' published in October 1928 issue of *Indian Journal of Economics*. See Sundar Ram, L. L. (1928) 'Indian Pastures and Fodder Supply', *Indian Journal of Economics*, Vol IX, Part 2, Serial No 33, pp.153–187.

10. Sundara Ram, L. L. (1926) 'The Indian Attitude towards Cow Protection', *The Maha-Bodhi*, Vol XXXIV, No 10, (October), pp. 497–506.
11. For more on the history of this monthly journal of the Maha Bodhi Society, see: https://mbsiindia.org/the-journal-of-the-maha-bodhi-society-its-origin-and-early-history/.
12. See, Mallik, S. N. (1929) Review of L. L. Sundara Rama's *Cow Protection in India*, in *Asiatic Review*, XXV, (January 1929), pp. 161–162.
13. Similarly, when an American friend requested Gandhiji to supply to him a comprehensive book outlining the cause of Cow Protection in India, Gandhiji wrote in Notes section of fortnightly, *Young India* that since "a text book on cow protection has been found to be a necessity" and since he failed to find for this American friend "a volume (on cow protection) that would give him all the information that he needed", Gandhiji then "approached Sjt. Ravashankar Jagjivan and asked him whether he would issue a prize for a cow-protection essay". On receiving consent from Revashankar Jagjivan, Gandhiji declared a prize for INR 1000 for the best essay submitted to All India Cow Protection Association to be received at AICPA office in Satyagrah Ashram, Sabarmati (Ahmedabad) on or before 31st March 1926. Gandhiji had invited Acharya Anandshankar Dhruva and Sjt. C. V. Vaidya to serve as judges of the essay competition. See, *The Collected Works of Mahatma Gandhi*, Vol XXVIII (August to November 1925), the Publications Division, Ministry of Information and Broadcasting, April 1968, pp. 390–391.
14. Puniyani, Ram (2017) 'Hindutva and Cow Agenda: Love for Cow or Hatred for "Others"'? *Mainstream*, Vol LV, No 30 (July 15, 2017). https://www.mainstreamweekly.net/article7308.html.
15. For the detailed discussion on the cattle numbers see, Low, C. E. (1912) 'The Supply of Agricultural Cattle in India', *The Agricultural Journal of India*, Vol VII, Part iv, pp. 331–342.
16. See, Woodroffe, John (1926) 'A Note on the Cattle of India: With Special Reference to Breeding, Pasturage, Slaughter and Exports', Reproduced in Chappterjee, Nilanda (1926) *The Condition of Cattle in India: Being an Inquiry into the Causes of the Present Deterioration of Cattle with Suggestions for their Remedy*, All India Cow Conference Association, Calcutta, pp. 13–15 in Appendices.
17. Third Humanitarian Conference that took place in Bombay had passed a resolution that read as: "That this conference respectfully appeals to the Government of India to prohibit by law the slaughter of milch and agricultural cattle, as well as of their young stock, as it deprives the Motherland of one of its valuable assets, and thus impairs the progress of agriculture on which depends the maintenance of two-thirds of its population and causes scarcity of the milk-supply and milk-products.". See, *Report of the Third Humanitarian Conference, Bombay held on 2nd and 3rd September 1918*, The Bombay Humanitarian League, Bombay, pp. 45–48.
18. Matson, J. (1933) 'Inefficiency of Cattle in India through Disease,' Agriculture and Livestock in India No 1, pp. 227–228.
19. Subsequent chapters in this book deal with the question in Bombay and the different drives to salvage dry cattle from Bombay city.
20. Harris, Marvin (1966) 'The Cultural Ecology of India's Sacred Cattle', Current Anthropology, Vol VII, No 1, pp. 51–59.
21. Knight, Henry (1954) *Food Administration in India 1939–1947*, Stanford University Press, Stanford.
22. Lensch, J. (1987) *Problems and Prospects of Cattle and Buffalo Husbandry in India*, Hamburg, Germany.
23. Trivedi, A. B. (1949) Post-War Gujarat: Economic Survey after World War II, A. B Trivedi, Khalsa College, Bombay, P. 76.
24. *The Indian Taxation Enquiry Committee* (1925), Vol III, Evidence Delhi and Benares, Central Publication Branch, Government of India, Calcutta.
25. Mukherjee, Radhakamal (1938) *Food Planning for Four Hundred Million*, Macmillan, London, p. 138; Cited in Raj, K. N. (1971) 'India's Sacred Cattle: Theories and Empirical Findings', *Economic and Political Weekly*, Vol 6, No 13 (27th March 1971), Pp. 717 + 719–722.

26. Sen, K. C. and S. C. Ray (1941) 'Animal Husbandry and Crop Planning in India', *Science and Culture*, Vol VI, No 2, (June 1941), pp. 684–689.
27. *Report on India's Food Crisis and Steps to Meet It*, The Agricultural Production Team (sponsored by The Ford Foundation), Ministry of Food and Agriculture, Government of India, April 1959.
28. Whyte, R. O. (1958) 'International Development of Grazing and Fodder Resources: Part V India,' Grass and Forage Science, Vol 13, Issue 3, pp. 203–212.
29. Chatterjee, I. (1962) 'A Single Species Cattle Economy', *Agricultural Situation in India*, Vol 17, No 7, pp. 764–784.
30. Dandekar, V. M. (1964) 'Problem of Numbers in Cattle Development', The Economic Weekly, (Annual Number February 1964), pp. 351–355. Whyte, R. O. (1967) Milk Production in Developing Countries, Faber and Faber, London. Heston, Alan (1971) 'An Approach to the Sacred Cow of India', Current Anthropology, Vol 12, pp. 191–197. Mishra, S. N. (1978) Livestock Planning in India, Vikas, New Delhi.
31. Right from mid 1930s, we witness number of authors writing on the potential of buffalo to emerge as a major milch animal in India.
32. Olver, A. and M. Vaidyanathan (1933) Assessment of the Annual Contribution of Livestock to Indian Economy, in Proceedings of the First meeting of Animal Husbandry wing of Board of Agriculture in India.
33. Wright, N. C. (1937) Report on the Development of the Cattle and Dairy Industries of India, Government of India Press, Shimla.
34. Emphasizing on the importance of establishing herd books and putting in place a rigorous milk records system, Wright (1937: 70) had pointed out that investigation into subjects such as lactation curve and the effect of age on subject's lactation yield holds more than mere academic relevance. While the work on lactation curve had been carried out in early 1930s by two animal husbandry experts, L. C. Sikka and K. R. P. Kartha, the establishment of a proper system of milk records had not yet been put into place. At the time of the second meeting of Animal Husbandry Wing of Board of Agriculture and Animal Husbandry (i.e. December 1936), the subcommittee on Dairying (1938: 34) had discussed the subject and felt that "the time is ripe for the formation of breed and milk recording societies as an aid to developing the dairy industry of the country". Also see, the Note prepared by Zal R. Kothawala, the then Imperial Dairy Expert, for the second meeting of Animal Husbandry wing (Appendix XXVII in the Proceedings of the Second Meeting of Animal Husbandry Wing, ICAR, New Delhi, 1938, pp. 185–187). The preliminary work by Imperial Council of Agriculture Research covered only six breeds of cattle, namely, Sahiwal, Scindi, Tharparkar, Hariana, Gir and Ongole and the Murrah buffalo.
35. This may suggest that a little more than half of the milk produced was not reserved for household or neighborhood consumption and the same was being marketed to satiate urban demand.
36. However, if we compare this with per capita milk and milk products production and consumption figures with the ones quoted in an authoritative 'Report on the Development of Cattle and Dairy Industries in India' by N C Wright (1937: 02), we come to realize that there is a huge difference between the two estimates, since Wright (1937: 02 and 155) puts it in the range between 7 to 8 oz per day! Philips, R. W. (1945) 'Impressions of India', The Scientific Monthly, Vol 61, No 5, (November 1945), pp. 397–412.
37. Kartha, K R P (1933) 'A Note on the Comparative Economic Efficiency of the Indian Cow, the Half-bred Cow and the Buffalo as producers of Milk and Butter-fat,' presented at First Meeting of Animal Husbandry Wing of Board of Agriculture and Animal Husbandry in India, ICAR, February 1933. Quoted in Wright (1937: 65).
38. Duckham, A. N. (1932) 'Animal Industry in British Empire', Oxford University Press, London. Quoted in Wright (1937: 57).
39. Walton, H. G. (1903) A Monograph on Training and Working in Leather in the United Provinces, Government Press, Allahabad, pp. 4–5.

40. After citing this reference, Dharampal extends the guestimate to suggest that "probably every district in British India had one or more slaughterhouses, which also carried out cow slaughter". However, this guestimate is not supported by advancing evidence or attempting to draw a geography of cattle slaughter. Dharampal seems to bring the practice of cow slaughter during the British rule of over 150 years (1800–1945) under immense scrutiny and heavily relies on a statement attributed to Mohandas Gandhi speaking in the year 1917 and quoting a figure of 30,000 cows getting slaughtered daily. Dharampal (2002) *The British Origin of Cow Slaughter in India: With Some British Documents on the Anti-Kine Killing Movement, 1880–1894*, Society for Integrated Development of the Himalayas, Massoorie.
41. With estimated 300 million acres under cultivable land status, as Wright (1937: 58) shows, the Royal Commission on Agriculture assumption itself would mean that India required 60 million bullocks, a figure that goes up when one takes into account a pair of bullock as basic unit on family farms whose average size was much below 10 acres in India of those years and hasn't gone up by a very high degree even as of now and has shown actually a downward trend.
42. In 1952, when N C Wright spoke at a Royal Society meeting on 'Some Problems of Tropical Dairying', he stressed that there needs to be more effort to push for mixed farming, but underlined the fact that since there is a shortage of fuel, which results in the use of dung as fuel, rather than as manure. Proceeding of the Royal Society on Dairying, 8th January 1952, Morning Session, pp. 149–157.

Chapter 3
Famines, Cattle Diseases and Cattle Mortalities: History of Civil Veterinary Departments in India

> The material culture of nineteenth- and twentieth-century Bengal revolved around domesticated animals. Agricultural cultivation, its mainstay of economy, depended primarily on cattle. The urban city life relied on draft animals which also featured in food and rituals of the indigenous Bengali subjects. Yet, in spite of the socio-economic implication of animals in Indian culture, contemporary South Asian scholarship is somewhat silent on the importance of studying non-human animals as historical actors.
>
> Samanta, Samiparna (2016) 'Cattle, Cruelty, Cow Doctors: Examining Animal Health in Rural Bengal, 1850-1920', in Kumar, Deepak and Bipasha Raha (eds) (2016) *Tilling the Land: Agricultural Knowledge and Practices in Colonial India*, Primus Books, Delhi. Pp. 214-237.

While such acknowledgement has come to arise within contemporary South Asian scholarship engaging with agrarian history, if one looks at the edited volume where Samanta articulates the above sentiment, we find only two articles out of thirteen there, devoted to the role played by bovine animals in India's agrarian and agro-pastoral landscape. This chapter presents how the bovine animals and their health demanded attention from the actors within late colonial state by paying close attention to cattle diseases, cattle mortalities and famines following in the footsteps of Saurabh Mishra and Samiparna Samanta.[1] This chapter weaves in the narrative of how the early thinking on addressing the cattle diseases and epizootics surfaced and alongside it arose proposals of establishing Civil Veterinary Department in the decade of 1860s. However, the late colonial state was extremely slow to respond to the concerns around saving 'public cattle', and even the inquiries on the cattle plague in Rawalpindi division in 1878–79 were partly motivated by militaristic motives, since the outbreak of cattle plague there had coincided with the Second Afghan War and had wiped out three fourth of cattle employed in transportation.[2] For the first 100 years, from 1793 to 1892, while Veterinary services and profession made an imprint, in India it was marked by an overwhelming preoccupation with the health of horses and other military animals. This preoccupation continued so much so that even when the first Civil Veterinary Department took shape in Bombay Presidency around 1887, for several years all that is reported in annual administrative reports was comments about

H. Upadhyaya, *Critical Insights on Colonial Modes of Seeing Cattle in India (1850–1980)*, Asia in Transition 27, https://doi.org/10.1007/978-981-97-1560-2_3

horse breeding operations and the phrase 'cattle disease' remains conspicuously absent in these administrative reports.

The proposal for creating the Civil Veterinary Department in India was mooted first in the year 1869, when the Commission on Cattle Plague in India was formed. The Commission presented its 1100 pages report in the year 1871. However, for two decades thereafter, the matter of creating the Civil Veterinary Department had remained perpetually "under consideration". The initial steps to form a Civil Veterinary Department were taken only in 1888. "Before 1888, four veterinary officers under the Military Department had been engaged in horse breeding operations to serve the Empire's militaristic interest. In 1892, it was decided to incorporate these officers in the Civil Veterinary Department, which was to consist, in the first instance, 18 officers recruited from the Army Veterinary Department".[3]

Similarly on the front of Veterinary Education, in 1862, the first Army Veterinary School was established in Poona (in Bombay Presidency) at the suggestion of Staff Veterinary Surgeon, J. H. B. Hallen and soon thereafter in 1877, the first Civil Veterinary School was proposed and organized by Inspecting Veterinary Surgeon, J. H. B. Hallen and the General Superintendent, Horse Breeding Operations for the purpose of educating men for the position of salutris and castrators in the horse and mule breeding districts of Northern India. However, five years later in 1882, a Veterinary College was established at Lahore, with a dispensary and a hospital attached to it, and the Veterinary School at Babugarh (Hapur) was closed down and 35 students were transferred to Lahore Veterinary School to continue their education.[4] In 1890, when Voelcker visited India, Veterinary school at Lahore had around 90 students.[5] Voelcker had also referred to a Veterinary training course being on the syllabus of Poona College and men who had passed it were considered "qualified to take charge of the local dispensaries which had been started at Ahmedabad, Nadiad and other towns in the Bombay Presidency".

The proposal for establishing a veterinary school at Ajmer was put forward by Col. G. H. Trevor, Agent to the Governor General in Rajputana around February 1891 and received warm support from Hallen, who had proposed a decade back in 1881 that "instruction in veterinary matters should be given to young native chiefs of the Mayo college at Ajmere (sic) in order that they might have some knowledge regarding the management, breeding and rearing of horses and cattle stock, so that when reigning in their several States they might take interest in such matters and induce their subject and tenantry to adopt the industries of cattle and horse breeding on scientific lines".[6] The annual administrative report of the Civil Veterinary Department of India for the year 1892–93 had reported that "the necessary buildings are now being erected on a suitable site in a suburb of Ajmere (sic); Veterinary Captain Hagger has been appointed Principal; a staff of lecturers will shortly be nominated and the school will be formally opened during the coming winter (i.e. winter of 1893)".[7]

During the summer of 1888, a veterinary class was conducted in Simla as suggested by Hallen and since it proved out to be useful, it was recommended that during the winter such a veterinary class shall be conducted at Umballa for non-commissioned officers, farriers, etc.

A veterinary college at Calcutta was proposed by the Cattle Plague Commission (1869–1871) "with a view to providing duly qualified native veterinarians for the cattle districts of Lower Bengal and Assam". While the proposal had received enthusiastic support from Earl Mayo and Hallen was given the task of selecting suitable premises for the proposed institution, two events derailed progress on this proposal. Following the assassination of Earl Mayo on 8th February 1872, the project was abandoned and immediately after working as President of the Cattle Plague Inquiry Commission (1869–1871), J. H. B. Hallen's services were requested on 16th December 1872, when he was put in the charge of newly appointed Special Studs Commission for the task of breaking the old Indian Studs (16th December 1872–31st March 1876). He devised the new scheme, in response to Lord Northbrook's policy "to breed horses and mules in India rather than import them from Australia and Persia, as was the previous practice", and during the period 1873–1894, supervised this work of reorganizing horse breeding operations on the military side.[8]

Thus, there was no further progress on the proposal of establishing Veterinary School at Calcutta until 1885, when a committee headed by Dr Kenneth McLeod was appointed to review the idea. McLeod committee submitted the report again recommending the establishment of Veterinary School at Calcutta and even described a site suitable for the purpose. However, no further action was taken until the year 1891, when Babu Sheo Baksh Bangla very generously subscribed and paid the Government Rs. 30,000 in aid of the project and pledged 7.25 bighas of land for the site. Similarly, Sir Dinshaw Manockji Petit, barrister of Bombay also came forward and promised a donation of Rs. 25,000 for the establishment of a hospital, attached to the institute. The annual administrative report of the Civil Veterinary Department of India for the year 1892–93 had reported that, "The committee for the organization of the school and hospital, with the assistance of Veterinary Captain W. D. Gunn (who has been appointed Principal), arranged and superintended the erection of the institution, which is now nearly completed. It is expected that it will be formally opened during the winter of 1893–94".

Similarly, in the year 1863, "J. H. B. Hallen suggested to the Government of Bombay the advisability of inaugurating a veterinary college and hospital at Bombay for the education of veterinarians in the civil districts of the Bombay Presidency". It appears that this proposal was rooted in response to "the heavy losses of cattle from contagious diseases" and an urge to "afford veterinary aid to the native agriculturist". However, sadly the proposal registered no further progress until the year 1886, when W. Lamb, principal veterinary surgeon, Bombay Presidency urged upon the government to establish a veterinary school at Bombay, and this was done. For this institute also generous donation was received from Sir Dinshaw Manockji Petit. On 26th May 1886, Professor J. H. Steel joined as its principal and the college opened with 69 students.[9]

As on 1892–93, there was also a Veterinary school in operation at Rangoon, Burma under the leadership of Veterinary Lieutenant G. H. Evans. This school used to deliver instructions in Burmese language and produced qualified graduates thoroughly trained in undertaking inspection duties and render their services in the suppressions of outbreaks of contagious diseases. Thus, as on 1892–93, when the first annual administrative report of Civil Veterinary Department in India was submitted, only Madras Presidency had remained without the promise of a veterinary college or school. On 5th May 1890, Hallen had pressed upon the need to establish a veterinary school at Madras in a letter sent to the Chief Secretary to the Government of Madras, arguing out his position that "the veterinary class attached to Saidapet Agriculture College, 8 or 9 miles from Madras, and so out of reach of owners of sick and lame animals in the city, has proved a failure in providing practical native veterinarians, as the practice in Veterinary hospital attached to the Agricultural College has been so very limited".[10] Hallen had kept pressing on this need for a veterinary school or college at Madras with follow-up and reminders, but unfortunately, in an order dated 20th October 1893, the Government of Madras stated, "The proposal to establish a veterinary school in Madras town does not in the present state of affairs commend itself to Government; the idea must, therefore, be abandoned".[11] In his first administrative report of Civil Veterinary Department in India for the year 1892–93, Hallen lamented that "*It appears to me strange that Madras Government have not determined to inaugurate a veterinary school in Madras...The losses sustained annually by the native agriculturist of Madras Presidency have been acknowledged as very heavy. Returns furnished a few years ago afforded evidence of this (heavy loss)*". (emphasis added).

However, before we discuss further the work of Civil Veterinary Department in India, let's look back into what existed on the front of veterinary services during the East India Company rule and briefly look into what motivated the early interventions around horse breeding operations taking place at Studs in the Bengal Army since 1793. In his book, *A History of Royal Army Veterinary Corps* (1796–1919), Major General Sir Frederick Smith (1927: 01–02) laments that not much was known about the role played by veterinarians working with Corps, as very few have left behind memoirs. Discussing the early years of Army Veterinary Service in India, he states:

> The Veterinary Service of the East India Company was established in 1826 and abolished in 1858. Thus, it took part in every epoch-making campaign, but the officers carried the information they acquired to their graves: not a syllable was written by any of them".[12]

Trying to reconstruct an account of Army Veterinary Services in India, including the work done by those in charge of Studs and horse breeding operations for the Bengal Army, Smith (1927: 96–97) states:

> The earliest official reference I have found in connection with the Veterinary service in India is contained in General Order of the Bengal Army, dated 8th April 1793, in which it is stated that upon the appearance of a supposed infectious disease among the horses, the affected will be removed from the cantonment, and the matter reported to the Officer Commanding the Station, who will order the animals to be inspected "by persons who may be judged capable of distinguishing the disorders of horses". We are left in doubt as to whether these were Europeans or natives. *At this time, there was not a single qualified man in India and only two in England, Sainbel and Moorcroft.* (emphasis added)

Another recent scholar Lt Col J. S. Bhalla who was commissioned to produce a similar work of attempting an institutional history that traces the history of the Remount and Veterinary Corps in India also picks up the date of 1794 as marking 'inception'. Mishra (2016: 28) while reflecting on Bhalla's choice of picking up this date (i.e. 1794) as marking 'inception' for remount operations in India further clarifies that, "A board for superintending over horse breeding activities (in India) was formed in 1794 and so if a date of inception for remount operations has to be chosen, perhaps this is as appropriate a date as any".

The above-mentioned General Order of the Bengal Army bearing the date 8th April 1793 and its contents makes us appreciate that two issues around war horses had occupied the attention of colonial state in India, the infectious diseases amongst the horses and horse breeding operations respectively around 1793 and 1794. The last sentence in the above quote from Smith's *History of Royal Army Veterinary Corps* also gives us enough hint to make a conjecture that veterinary medicine and science was in its infancy around that period and those who were supposed to implement those directions in the event of infectious disease amongst the horses would have been amateur animal health practitioners, relying on the traditional healing practices handed over from generation to generation.

Smith (1927: 97) gives an account of the work carried out under the direction and leadership of Moorcroft, who arrived in India on 14th November 1808 and points out that "he gave up a lucrative practice (in England) to become Superintendent of Studs, and his duties were to reorganize the Breeding Studs of the Bengal". Smith (1927: 97) tells us that in his 11-year-long career in India, Moorcroft "had trained a doctor as an assistant (Surgeon Gibb) and had made great progress in the suppression of the contagious diseases, demonstrating what veterinary science could accomplish in this direction". However, there comes a dramatic turn of events in 1819, when Moorcroft undertakes journey to Central Asia in order to obtain the animals since "he had rejected the Arab and English horses as not producing the type at which he was aiming" and believed that Central Asia may offer him such promising animal that he was looking for. We are told by Smith (1927: 97–98) that for next seven years from the time of Moorcroft's departure, "the Government of Bengal had kept his appointment open, with Dr. Gibb in charge, but as soon as definite tidings of his death were obtained, Moorcroft's breeding establishments were, in 1826 returned to the control of their original incompetent superintendents, inexperienced infantry and cavalry officers, and Dr. Gibb disappeared".

The story of the Stud Department from that moment for the next half century (i.e. 1826 to 1873) was full of failures and terrible losses. Drawing upon the documents and correspondence connected with the Indian Studs from 1793 to 1873 at the India Office, Smith (1927: 98) summarizes this dramatic period in the following words:

> This is not the place to give an account of the fortunes of the Studs during the succeeding half century. The product was the laughingstock of the Army and the despair of the East India Company. Not once or twice after a clean sweep of the various establishments was a fresh start made, but always under the same incompetent management. From first to last, the Studs cost millions; the mismanagement was staggering and unbelievable, and the story of the repeated costly failures reads like a page of extravagant fiction.

3.1 Cattle Plague and Early Inquiries into Epizootics

As remarked earlier, murrains amongst cattle in the decade of 1840s in Madras and then in 1860s in Calcutta and in late 1870s in Rawalpindi division in Punjab had pressed late colonial state to initiate inquiries, however, those inquiries did not lead to protracted action on providing protection to public cattle against contagious diseases.

> In March last, I received intimation that I had been placed on special duty under Punjab Government, for the purpose of investigating and devising some remedial measures for a disease which had been and still was very prevalent amongst cattle in Rawalpindi division. The first steps necessary to be taken were those which would throw some light on the nature of the disease, and would thereby enable me to recommend measures either sanitary or curative. With this object in view, I visited the commissariat cattle, which were then suffering to a great extent, and I also inspected various villages within a circuit of fifteen miles from Rawalpindi. I had not to proceed far with these investigations, before I felt fully convinced that the outbreak which had to be combated was that known as Cattle Plague. *These visits, however, disclosed other facts, the most important of which was that I had to deal with a people totally ignorant of the simplest laws of sanitation, too apathetic to watch their own interests, and consequently too indifferent to consider for one moment those of their neighbours. What then was the state of affairs found to exist? On entering a village, the lumbardar, or headman, invariably reported about 50 to 75 percent of the cattle to have died. On every side the zamindars, or small farmers, were to be seen idling, having in many cases no bullocks left with which they could till their land.* (emphasis added)

The above quote occurs in a paper by A. E. Queripel titled 'Cattle plague in India' that was published in *The Veterinary Journal and Annals of Comparative Pathology* (Vol IX, Issue 11, November 1879, Pp. 310–316). Giving readers a glimpse of cattle mortalities arising from Cattle Plague in Rawalpindi division, Queripel (1879: 314) wrote:

> From observations made in various localities, I would say that the zamindars have alone lost 60,000 whilst the loss sustained by the various contractors of the commissariat train can not be computed at less than 15,000. The government bullock train, employing 1200 head of cattle, sustained a loss of 800; while again the slaughter animals belonging to the commissariat suffered greatly. It must be remembered that the above only represent the losses of this division; the disease has been very widely spread in the Peshawur (sic) district, and when in Wazirabad, I found that nearly 5000 had died from the fatal malady.

Same as "Murrain of Europe" was how the *Burra azar* of Madras was talked about by Dr Gilchrist in 1848, and Dr. C. Palmer showed no hesitation in 1864, to call the "epizootic of Calcutta" same as Rinderpest being witnessed in Europe then. Cattle Plague Commission added to these two earlier observations, the opinion pronounced by Mr. Gudgin, Veterinary Surgeon in Burma, who after carefully studying the epizootic in Burma, concluded that it was same as Rinderpest. The Report of the Cattle Plague Commission (1871: vii–viii) makes an interesting observation on the change in attitude towards cattle rearing affected by the promise of wealth creation in presidency towns such as Calcutta stating that "As the surplus and unproductive stock of the country becomes less, as cattle becomes less a representative of wealth, and more an agency of wealth, the subject of losses by murrain will assume a greater importance and attract a closer attention and interest, and the necessity for repressive measures becomes more declared and pressing".

In the year 1872, following the voluminous report by the Cattle Plague Commission in India which ran into 1100 pages, *The Calcutta Review* carried a longer article titled 'The Indian Rinderpest'.[13] This longer article not only summarized the Cattle Plague Commission report, but reflecting upon lukewarm and passing interest shown in the report on such an important subject by press in India with a shock, the article (1872: 169) stated:

> On what principles do general readers in India, who interest themselves in public affairs, assess the relative value of the various subjects which are from time to time brought under their notice?...We cannot tell; we can only say, that very frequently it is now as in ancient times, - 'the race is not to the swift nor the battle to the strong'; solid labour for the general and present good is often disregarded, and its record is unread, while every journalist vies with his brethren in reproducing and criticizing any pleasing translation of ancient Hindu fable. These remarks have been naturally suggested to us by observing the almost total neglect which the learned and copious labours of the Cattle Plague Commission have met with at the hands of Indian Press. We are well aware that a notice of the Report as published, has been given by most of the Indian journals; but these, as a rule, have gone no deeper into the matter than an entry in a catalogue might be expected to go, and prove only one thing with perfect clearness, that those who wrote the notices – one and all – knew little or nothing of the subject of cattle plague when they took up the report, and failed from such perusal as they gave that report, to cull from it any fact which could be of service in instructing the general public. Some even have contented themselves with describing mainly the outside of the book, making silly little jokes based on its weight and size.

3.2 Civil Veterinary Department in Bombay Presidency and Reporting on Cattle Diseases

In late 1880s, Bombay Presidency became the first presidency to have its own provincial Civil Veterinary Department, but the work performed remained around the horse breeding operations in the region.[14] Shortly thereafter, in 1890, the Imperial Bacteriological Laboratory was started in Poona under the Civil Veterinary Department and the services of Dr. Alfred Lingard were secured in order to take up bacteriological inquiries on contagious diseases of cattle. However, due to climatic conditions, the premises at Poona proved out to be unsuitable and after two years of experience, it was decided to shift the location of bacteriological laboratory to more suitable site in Kumaon hills.

While the initial action on organizing Civil Veterinary Department in India appeared in January 1892, it is very interesting to study the way, *The Veterinary Record* (a weekly journal of the profession published in Britain and edited by Willian Hunting) voiced the editor's remarks on the first circular issued by the War Office in January 1892 calling for volunteers to apply for "continuous service" in India for the proposed Civil Veterinary Department and to voice willingness to be thus recruited by return post. The editorial stated:

A circular has been issued by the War Office calling for volunteers to join the Indian Veterinary Department. The candidates will be bound for "continuous service" in India, and will be called upon to resume military duty in the event of serious war – whatever that means. The circular goes on to say that the Civil Department will receive the same pay as the military plus a staff allowance. So far good, but we wish to draw the attention of intending candidates to the fact that nothing is said about ordinary leave, sick leave, retiring allowance &c. They will get a staff allowance, and will serve continuously in India. These omissions suggest caution. *To be an uncovenanted Indian civilian is not the happiest way of making provision for the future, and the Government of India enjoys a very evil reputation in regard to its dealings with the Veterinary Department. On this account it is desirable that all important points should be settled in black and white before anyone volunteers for what is perpetual exile, according to the circular. It is not our desire to prevent officers joining this department, but we advise them to take such ordinary precautions as are laid down in any respectable emigrant office. Holding these views, it seems to us a little hasty to issue a call for volunteers by 'return of post'.*[15] (emphasis added)

Similar concerns and cautionary advice were voiced in the newspaper published from Madras Presidency a month later in its issue dated 20th February 1892, which in turn was picked up by the weekly journal, *The Veterinary Record* in its issue dated 26th March 1892. After commenting on its displeasure on the salary offered to those 18 officers who were to be hired for the proposed Civil Veterinary Department in India, the newspaper editorial stated:

Out of eighteen Veterinary Surgeons who were to have formed the Civil Veterinary Department, four were for remount service, and four for horse breeding, so that only ten remained for the poor agriculturist, and these were to be scattered over the whole of India and Burma. *The postponement of so important a department in the face of the yearly losses to agricultural aid alone is a serious mistake. The yearly statistics, and those for the past year in particular, should open the eyes of the Government to the need of a Civil Veterinary Department regardless of what it may cost. The wealth of a nation depends much upon its population and agriculture, and without the bullock to plough and the cow to produce the bullock, agriculture will wane, and wealth and population suffer".*[16] (emphasis added)

It becomes clear on the close examination of correspondence and documents that while the Civil Veterinary Department was supposed to be established in January 1892, it was not until July 1893 that rules were finally issued for acceptance. A resolution of the Government of India, Revenue and Agriculture, No. 26–123, dated 4th July 1893 had laid down the precise duties of Civil Veterinary Department, with having the followings as the main heads: (i) Horse Breeding, (ii) Animal disease and (iii) Instruction. This was followed by another circular from Revenue and Agriculture Department, Government of India, No. 41–47, issued in October 1893 that pointed out "the necessity for organizing a subordinate establishment under each Local Government".

In 1894, Civil Veterinary Department published five ledgers on the following subjects: (i) Rinderpest (Cattle Plague), (ii) Anthrax, (iii) Charbon Symptomatique of the Ox, (iv) Horse Surra and (v) Repression and Prevention of Cattle Diseases in India. The ledger on Rinderpest cites an example of the disease spreading though the pack animals from the Kulu valley in the Punjab into Lahoul in 1890, and it had attacked the yaks and their progeny there in all the villages to Kyelang, leaving but one or two animals alive in its track.

In 1904, there was issued a very important dispatch, No. 137, bearing the date 5th May, which stated that "an increased number of (veterinary) assistants of necessity entails the appointment of additional inspecting officers *if the best results are to be obtained and if money is not to be wasted wholesale for want of effective supervision*". The dispatch had a tone of sympathy when it noted that they were "convinced that to stint the superior staff unduly was the worst political economy". This dispatch had also addressed the problem faced by many Superintendents of Provincial Civil Veterinary Departments who had a very expansive terrain to cover and stated, "it must not be overlooked that, when an expensive officer has too large a charge, an excessive portion of his time is spent in travelling which should be devoted to inspection". At the time of dispatch, there were 15 Veterinary Inspectors and 307 Veterinary Assistants at work in India. Twelve years period following this dispatch witnessed a steady expansion of the Civil Veterinary Department in India and in 1916, there were 119 Veterinary Inspectors and 1056 Veterinary Assistants at work in India.

In 1908, there was a meeting of Civil Veterinary Department Officers at Calcutta. At this time, Veterinary Conference was independent of the Board of Agriculture in India, and it was resolved at this conference that a separate Board of Veterinary Science be constituted, but the proposal had not materialized. However, there was one grave episode that had far-reaching consequences for the Civil Veterinary Department of India. This event occurred in April 1912 when the Inspector General position that provided leadership and direction to Civil Veterinary Department was abolished. More intense historical scholarship is required to understand whether the Civil Veterinary Department for the longest period of time was a service department that was made subservient to the more prestigious and powerful Department of Agriculture. The existing narrative from Saurabh Mishra suggests that the veterinary work happening in British India till the date of 1860s had remained excessively interested in the health of military animals and with operations such as horse breeding, at the cost of ignoring 'public cattle'.[17]

The First World War years proved to be especially ominous for the veterinary department personnel serving in different provinces. The wartime emergencies had made demands on human resources from India's Civil Veterinary Department and "out of 28 European officers, 10 were entirely released for military duty".[18] The direct influence of the war had been the total closing down of recruiting in Civil Veterinary Department of India on one side and on the other had demanded human resources from the department to take up military duty, thereby impacting a reduction in the department at both the superior and subordinate grades to the lowest possible working scale. In his opening remarks at the First Meeting of Veterinary Officers in India in March 1919, Mckena also voiced his concern that there existed "the utter inadequacy of the veterinary staff, either in the superior or subordinate grades to cope with the vast problems with which the department is confronted".[19] Making an ardent plea for a large and an early expansion of the Civil Veterinary Department, Mckena emphatically stated his concern by saying, "[O]n the maintenance and health of the cattle of India depends, the whole of our agriculture and, if that is to be effectively achieved, it seems necessary that we should have a large and an early expansion of our department".[20]

Thus, at the first meeting of veterinary officers in India in the year 1919, there was an overwhelming concern that the Civil Veterinary Department had got the short end of the stick. One of the several resolutions passed at the meeting expressed this viewpoint without mincing words:

> This meeting is of opinion that *the present constitution and organization of the Civil Veterinary Department are utterly inadequate to cope with the work demanded of it,* and it, therefore, respectfully represents to the Government of India the absolute necessity of filling up all existing vacancies in the Indian Civil Veterinary Department without delay and would emphasise the importance of a large expansion of all provincial work of the country.

The frustration that had sunk in the veterinary officers by 1919 also got expressed when Mr. Taylor was speaking out during the discussion on the proposed revival of the 'Journal of Tropical Veterinary Science'. He stated in a somewhat desolate and angry mood that "*with the present depleted staff and lack of facilities for original research,* it would be difficult to obtain material for a journal".

Colonel Walker, speaking at the first meeting of Veterinary Officers had said in a somewhat angry tone:

> I think I have said enough to show that the Imperial branch of the department is practically moribund, and that if it is intended to revive it, a reorganization is required to include the reappointment of a head. If nothing is contemplated, it were better to put the corpse in the coffin and give it a decent funeral, as soon as possible, (rather than) leaving it to Local Governments to appoint an officer occasionally in the same way as they might a Boiler Inspector.

3.3 Reporting Cattle Diseases

In the annual administrative report of the Civil Veterinary Department of Bombay Presidency for the year 1894–95, we find for the first time some remarks on cattle diseases. While reviewing the annual report, C. S. Curtis, Under Secretary to the Government observed:

> During the year under report (i.e. 1893-94) cattle disease prevailed more or less in all the districts of the Presidency proper except Panch Mahals and Dharwar, and in the district of Shikarpur in Sind. No reports were received by the superintendent from the Collectors of Khandesh, Surat and Bijapur, but the Survey Commissioner and Director has supplied the figures for the last two districts. Out of a total of 17,273 cattle reported to have been attacked by various contagious and infectious diseases 10,014 or 58 percent died. Disease was especially prevalent in the districts of Kanara and Sholapur where the figures stood as follows:
>
> Kanara: 7070 cases and 4178 deaths
>
> Sholapur: 2657 cases and 1246 deaths.
>
> The usual precautionary measures to prevent spread of the disease were employed by the Veterinary Assistants in those districts in which they are employed, but the Superintendent states that great difficulty is experienced by these officers in inducing the owners to segregate the infected cattle.[21]

Similarly, review remarks on the next year's annual administrative report showed that outbreaks of cattle diseases continued unabated:

> During the year under report (i.e. 1895-96) outbreaks of cattle disease were reported from 15 districts of Presidency, but as remarked by Veterinary Captain Morgan and Colonel Godfrey, other districts could not have been entirely free from it. The total number of animals attacked by various contagious and infectious diseases rose from 17,273 in 1894-95 to 18,635 in 1895-96, but the percentage of fatal cases decreased from 58 to 53.65. As many as 91.3 percent of the animals attacked suffered from rinderpest, and 92 percent of the total number of deaths were due to that disease. The usual precautionary measures to prevent the spread of the diseases were adopted by Veterinary Assistants in those districts in which they are employed.

Similar remarks showing the continuation of outbreaks of cattle disease in 17 districts of Presidency are noticed in the review of the annual administrative report of Civil Veterinary Department of Bombay Presidency for the year 1896–97:

> During 1896-97, outbreaks of cattle disease were reported from 17 districts of the Presidency, but no reports on the subject have been received from the districts of Panch Mahals, Dharwar, Hyderabad and Thar and Parkar for the last four years and from Surat for the last two years. As remarked by the Superintendent and the Survey Commissioner and Director, it is unlikely that these districts were free from disease and the attention of collectors should be drawn to this point. It appears that in several cases, veterinary graduates received information of outbreaks of cattle disease too late to enable them to arrive in time to do much good, and arrangements should therefore be made to communicate the information to the officers concerned more promptly in future. The total number of animals reported to have been attacked by various contagious and infectious diseases during the year under report was 18,513 against 18635 in 1895-96 and the percentage of fatal cases also fell from 53.65 to 47.58. Rinderpest accounted for 16,862 or 91 percent of the total number of attacks reported and 8,552 or 97 percent of total fatal cases. As usual, Kanara suffered very heavily from the disease and the Superintendent, Civil Veterinary Department should be requested to notice in his next report the results of the measures suggested by the Survey Commissioner and Director in paragraph 21 in his review for it suppression and prevention in that district.

Similar remarks showing the continuation of outbreaks of cattle disease in 19 districts of the Presidency are noticed in the review of the annual administrative report of Civil Veterinary Department of Bombay Presidency for the year 1897–98:

> During the year outbreaks of cattle disease were reported from 19 districts of the Presidency, of which 11 had the advantage of the services of Veterinary graduates, while no reports on the subject were received by the Superintendent from the districts of Sholapur, Surat, Hyderabad and Thar and Parkar. The total number of animals said to have been attacked by various diseases and the number of fatal cases amounted to 12,648 and 5918 in 1897-98 as against 20,713 and 10,211 (inclusive of 2200 attacks and 1400 deaths reported after the close of the year 1896-97). Rinderpest alone was responsible for 7,295 and 4,906 or nearly 58 percent and 83 percent respectively of the total number of attacks and deaths reported, the districts which suffered most heavily from this disease being Upper Sind Frontier, Shikarpur and Thana. It is a pity that the District Board of Kanara still find themselves unable to entertain even temporarily the services of a Veterinary graduate for the district.

On the cattle disease during 1899–1900, the review on annual administrative report of Bombay Presidency made the following remarks:

> Cattle disease was prevalent throughout the Presidency, and it is very doubtful whether the four districts reported free were really so. The collectors of these districts should see that the reports are carefully compiled. The high mortality which accompanied the outbreaks of rinderpest, causing 1338 out of the 1934 deaths during the year attributed to the epizootic disease, was probably, as observed by the Superintendent, often due to emaciation caused by famine, and in part to deaths from starvation or improper feeding being ascribed to that disease. Government have read with great interest the conclusion drawn by Veterinary Major Morgan from the small spread of disease in Ahmedabad camps. The report of the experiments referred to in paragraph 43 and 44 of this report will be awaited.

On the cattle disease during 1900–1901, the review on annual administrative report of Bombay Presidency made the following remarks:

> No cattle disease was reported from Broach, Surat, Khandesh, Ahmednagar, Sholapur, Bijapur, Thar and Parkar and Upper Sind Frontier. In the Presidency proper, 3136 animals were attacked against 5466 during the preceding year. Rinderpest claimed 68.9 percent of the attacks, foot and mouth disease 22.1 percent, anthrax, pleuro-pneumonia and carbon-symptomatique together 3.4 percent and other diseases 5.6 percent. Nashik suffered most. 54.30 percent of total number of attacks proved fatal.

Around this time in Bombay Presidency, we observe that the subject of inoculation against rinderpest starts getting mentioned. While the Superintendent is eager to undertake experiments on inoculation, these are not widespread in nature. The review states:

> No great progress has been made with the experiments with inoculation against rinderpest, undertaken by the Superintendent, Civil Veterinary Department. Bombay, partly owing to the delay in the receipt of the reports of outbreaks and partly to the sporadic nature of outbreaks. Twenty-seven villages were visited and 177 cattle successfully inoculated.

When Major Morgan had put up a cattle breeding scheme in Bombay Presidency in 1900, one commissioner had commented that "nothing more impresses a district officer than the immense good the Veterinary department may do to the country, *but I doubt if cattle breeding is the likeliest direction. I would rather ask them to concentrate more force against cattle disease*". This clearly shows the amount of investment and attention that cattle breeding tasks and policies received from the Civil Veterinary Department in Bombay Presidency. Part of the problem seems to arise from a vexed question: 'whether a veterinary officer or an agricultural officer is the right man to look after cattle breeding'.

In 1903, the horse breeding operation was taken over by the Army Remount Department in Bombay Presidency. However, within seven years, the task was handed back to the Civil Veterinary Department, in 1910. The war had also impacted the work of Major Hewlett who during those years, "engaged on military work, in addition to his own duties as Principal of the Bombay Veterinary College".[22]

3.4 Unsatisfactory State of Statistics of Cattle Mortalities (1890–1920s)

Although the section on 'cattle diseases' in Voelcker's report occupies just two pages, he very clearly notes that he would have liked to see more committed action "to learn more about the epidemics which annually clear off so many cattle of the country". Voelcker had advocated for the spread of veterinary education and dispensaries and had considered the appointment of Dr. Lingard to the post of Imperial Officer for Bacteriological Research in 1890 as "one of great importance" and "almost the first in which a man trained in scientific investigation has been brought to India and enabled to follow original research". Alluding to inquiries on cattle diseases carried out in Central Provinces and the state of Anantpur, he reported that "lakhs of rupees are annually lost by cattle diseases".[23]

Civil Veterinary Department had initiated the process of collecting and collating statistics on cattle mortality arising from contagious diseases in the year 1893–'94 and the inspector general noted the following words on the state of affairs on the front of 'cattle diseases—investigation and prevention' in the annual administrative report for the year:

> Under this heading but little actual progress has been made by the Civil Veterinary Department. Much information has been collected, but it has not as yet been tabulated.

Having summarized what reports from superintendents from different provinces suggested, the inspector general remarked on the enormity of the loss suffered due to cattle death from contagious diseases, even at a very conservative price being assigned to each head of cattle lost:

> The vast mortality shown to have occurred proves the enormous losses which the agricultural community suffers annually and shows the urgent necessity for the formation of our veterinary subordinate staff. Let us, roughly speaking, value each head of cattle lost at Rs 10, which is a very low valuation, and the loss from such cases as can be proved amounts to a large sum of Rs 7,48,570.

He also agreed with a complaining tone voiced by the superintendent, North Punjab saying that the horse breeding duties assigned to him had kept him so busy that there was practically very little time that could be devoted to investigation and prevention of cattle diseases. Similar hurdles existed in North–Western Provinces, Oudh and Bombay, where a large part of the time of superintendents was occupied in overseeing horse and mule breeding operations and he felt that this would necessitate the appointment of at least three more officers:

> I regret that more progress has not been made in this direction during the year under review; still *it must be remembered that the cattle portion of our work is in its infancy*, and that it will require two to three more years to mature.[24]

Expressing his views on how much work was done by the nascent Civil Veterinary Department of India on the front of cattle diseases, the inspector general stated:

> A start, but only a start, has been made in this direction. Information as to the prevalence of diseases of contagious nature has in some provinces been carefully collected, *but the majority of superintendents have not paid sufficient attention to this important subject.*

Unreliability of the statistics reported on the cattle diseases remained an issue that was stressed upon by the inspector general repeatedly in annual administrative reports of the Civil Veterinary Department of India. In his report for the year 1894–95, he had stated that:

> The statistics are, in the majority of the cases, absolutely unreliable and misleading. My own experience of the reports received from districts, is that they are practically valueless, and the number of deaths would have to be multiplied by many thousands to reach the true total.

In the following year's report (1895–'96), he quoted his remarks once more and stated that "the reports submitted this year only too fully confirm, what was said on the subject.[25] However, one can gauge enormity of the cattle mortality from contagious diseases in Bengal from what the Superintendent, Civil Veterinary Department stated in his annual report for the year 1895–'96:

> Since 1882, when Mr. Finucane, the then Director of Land Records and Agriculture, Bengal pointed out that losses from cattle diseases in Bengal amounted to a great deal more than eight lakhs per annum, the subject has been kept in view by Government, and my predecessor accumulated a good deal of information.[26]

Inspector General once again expressed his views that provincial superintendents of veterinary departments needed to devote themselves as much as possible to investigation of cattle disease and that their duties are or should be mainly connected with cattle:

> That something has been done in this direction (investigation and prevention of cattle diseases) during the year under review cannot be denied; that as much as I would have liked to have been able to report as having been accomplished, has not been done, I cannot deny; still with this small staff at our disposal and the Subordinate Establishment not under proper supervision and control, I feel confident that more could not be expected. There are, however, two Officers who could assist more in this direction. I allude to Superintendents, Civil Veterinary Departments, Central Provinces and Madras; their duties are or should be mainly connected with cattle, and they should devote themselves as much as possible to the investigation of disease: there cannot be a more important subject, and these Officers especially should do all that is in their power to throw light on obscure diseases in their provinces.[27]

He had expected that the Superintendent of the Civil Veterinary Department of Bombay should now have more time to devote to this subject, since he had been relieved of the charge of Sind in the year 1896–'97. He was of the opinion that excellent work in this direction was now being reported from Burma and Bengal and repeated his demand to get more Officers in other provinces and pleaded for an early decision on the Subordinate Veterinary Establishment.

3.5 Cattle Mortality During Famines in Bombay Presidency and Elsewhere

Statistics on cattle mortality were considered inexact and non-reliable not merely in the case of loss of cattle due to contagious disease. The same appeared to be true when it came to the estimation of loss of bovine stock due to famine conditions around this period, if these were compiled from the returns received from local revenue officials, such as *patwaris*, who were thought of as "inefficient, lazy and corrupt".[28] Instead of relying on these returns furnished by local revenue officials, veterinary and Agriculture Department officers considered it closer to the truth to arrive at estimation by extrapolating the data on the loss of bovine stock by looking at the export of hides from different provinces.[29]

A report on Famine in Punjab during two-year period of 1895–1897 stated that "it will probably never be known to what extent agricultural stock has been depleted during the past two years". It was felt that amongst all the Punjab districts, Hissar had faced considerable loss of cattle and one estimate had put the number of plough cattle lost there during 11 months between 1st May 1896 and 31st March 1897 at 77,134 or 54% of the agricultural stock that existed before the famine struck. Another estimate about loss of cattle in Hissar district stated, "1,32,325 animals, or 92% of the plough and well cattle of those found to exist in April 1895, disappeared in the two and a half years following".[30]

During the year 1899–1900, the great famine (*Chhappaniyo*) devastated the rural agrarian economy in many parts of western India and mainly in Gujarat. Veterinary major Morgan's report was quoted *in extenso* in the annual administrative report of the Civil Veterinary Department for the year 1899–1900:

> It is difficult to estimate the mortality which has taken place in certain parts of this (i.e. Bombay) Presidency. But there can be no doubt that the districts of Ahmedabad, Kaira, Punch Mahals and Broach have lost something like 70 percent. Surat being well supplied with fodder, both from grasslands and arable fields, has not suffered to the same extent. Before the famine, the above four districts had a little over 13 lakhs of head of cattle, and during the eight months, from August to March, exported about 300,000 mds of hides. Assuming six hides go to an Indian maund, the above figure represents a mortality, amounting to 1,800,000 head of cattle.[31]

The great famine also compelled the empire to pay some attention to the issue of 'preservation of cattle'. Report on the Indian Famine Commission (1901) notes:

> The great mortality of cattle in the recent famine has pushed to the front the question of their preservation in times of drought and dearth of fodder. Such fodder famines are fortunately rare. In an ordinary famine, when the crops fail at a last stage of their growth, there usually remains sufficient straw and grass to save, at any rate, the useful cattle; but the recent famine has been abnormal in this respect. It is estimated that nearly two million cattle, local and immigrant combined, died in the Central Provinces and its Feudatory States; and that an equal number died in Bombay. The mortality was also great in Berar and Ajmer, in which latter district no effective measures were taken to prevent it.[32]

The aftereffects of the devastating famine 'Chhappaniyo' (1899) had also cast its long shadow on the Northcote Gowshala managed by the Gujarat Cattle Preservation

Association. Mr. F. S. P. Lely, Commissioner, Northern Division in the annual report on *Cultural, Dairying and Breeding Operations conducted on Experimental Farms and Gardens in Bombay* talked about Northcote Farm experiencing "anxious time for the management owing to the withholding of rains". During the year 1901–'02, the farm was successful in selling 79 castrated male calves and as many as 50 were purchased by His Excellency the Governor with a view to getting them distributed through Mr. Lely "among some of the Patidars who have been hit hard by the famine".[33]

In the annual administrative report for the Civil Veterinary Department in the Bombay Presidency for year 1901–'02, the Superintendent voiced his opinion that "efforts should rather be directed to the preservation and increase of such special breeds of cattle as yet exist in certain localities in the Presidency than to the improvement of non-de-script herds such as are to be found in the Deccan".[34]

The year of the great famine *chhappaniyo*, 1899 also witnessed some steps taken by the Bombay government "to endeavour to minimize the excessive mortality arising from famine and to save the plough cattle of the cultivators". The Superintendent, Civil Veterinary Department recounted these efforts in his annual report thus:

> In the month of August 1899, the Government of Bombay realized that the famine in Guzerat had assumed such a serious aspect that the mortality amongst cattle would be so great, that it was necessary that some steps should be taken with a view of endeavouring to save even a small proportion in that district, and the experiment of establishing the cattle camps under European supervision was attempted. Large number of cattle were also sent by train to districts where grass and water were abundant.[35]

The cattle camps in eight places in Bombay Presidency accommodated around 7000 animals since October 1899 and upto the end of April 1900, the total casualties from debility (famine), accidents and common diseases amounted to 615 and from Rinderpest to only 68.[36] While this was an intervention that Civil Veterinary Department can take pride in, the experiment of sending cattle belonging to Ahmedabad district by rail to forest areas in Thana and Dharampur proved out to be failure. Morgan notes about this experiment in his annual report in these words:

> In month of October (1899), some 3500 cattle were railed from the Ahmedabad district and sent to the Thana forests and Dharampur. They were sent inland some 30 miles and grazed in the forests and uplands of the Thana district. The mortality was heavy, and I visited the various herds in April. No disease was apparent, and they were dying of starvation, as the grass was perfectly dry and yielded absolutely no nourishment and the water supply was indifferent. This somewhat doubtful experiment affords another instance of how careful one should be in moving cattle from one part of India to another either for breeding or any other purpose and how little adapted cattle of one district are to accommodate themselves to totally altered conditions.[37]

Another contemporary account estimates that 15,749 head of cattle were dispatched from Gujarat, of which 14,955 reached the grazing ground. The review of this experience estimates that 1065 cattle strayed while on their way to Thana forests and another 8731 cattle died and hence suggests that only around 5160 cattle out of 15,749 cattle dispatched had survived at the end of this experiment. It however also hints that a considerable number of cattle was brought back to Gujarat at various times by owners without the intervention or knowledge of controlling officers and hence total saved cattle may perhaps be more correctly recorded as about 7000.[38]

Report of the Indian Famine Commission (1901) candidly noted that the experiments made in the direction of the deportation of cattle to forests "cannot be said to have been successful". While 60 percent of those Gujarat cattle sent to Thana forests died, an estimated two-thirds of those sent to Berar forests perished.[39]

Even the policy of attempting to preserve cattle mortalities through the intervention of opening of cattle camps was deeply rooted in Malthusian impulses and hence there was a pronounced bias on which type of cattle would be admitted and how these cattle camps must be administered.

Such a selection bias got manifested at Ahmedabad Cattle Camp (opened in August 1899 and initially managed by Mr. Lambert, Superintendent of Police), whose management was handed over to Mr. Mollison and his subordinates of the Agriculture Department on 20th November 1899. As soon as the management was changed, a decision was taken to weed out all the animals not considered worth keeping and these 'worthless' cattle were sent to the Pinjrapol. "Their places were filled up by sound and useful stock, and the camp on regular and economic lines played a useful part throughout the year", noted the Commissioner, Northern Division, in his review of Cattle Camp Scheme. Who were the animals that got weeded out and who owned them?

> About a thousand animals were collected (at the first cattle camp opened by private subscriptions near the Shahibag in Ahmedabad) but mortality among them soon became excessive. They had not been selected by an expert and many of them were of inferior class. They were also unhealthy, *being mostly taken from the droves of Sindhi and other wandering graziers which had felt the pinch of scarcity for two years.*[40] (emphasis added)

Mishra (2015: 82–83) shows how a certain idea of not feeling compelled to preserve 'useless' cattle prevailed amongst colonial officials and hence "officials felt extremely exasperated with 'native philanthropist', who stubbornly continued their practice of feeding and protecting all kinds of cattle, thus obstructing the course of 'natural laws'". In his review of the Cattle Camp Scheme, the Commissioner, Northern Division had talked about the lessons from the experience of cattle camps and recommended that:

> Admissions should be made by an expert according to a high standard of breed, age and health. Animals that are debilitated cannot be restored or only at the cost of trouble and money that can ill be spared. Animals that have been fed largely on leaves or other green food cannot stand the change to dry grass.

Similarly, while writing down recommendations, Indian Famine Commission (1901) tried to rationalize this selection bias and remarked on the cattle camps thus:

> As the accommodation of a camp is limited and great risks are involved in overcrowding, *it is necessary to carefully select the cattle, and we think that admission should generally be restricted to selected cows with a few bulls of the best breed.* Experience suggests that it is better for Government to buy the cattle and re-sell them at a fair price, to their former owners, if possible, at the end of the famine, than to keep them at the risk and at the cost of the owners.[41]

Table 3.1 Impact of the Great Famine on Numbers of Bovine Stock in Ahmedabad (1876 to 1902)

	1876–'77	1880–'81	1890–'91	1900–'01	1901–'02
Plough Cattle	1,48,399	1,60,237	1,45,274	80,501	90,882
Milch Cattle	2,00,503	1,77,837	1,87,483	50,258	63,096
Breeding Bulls			1410	325	579
Oxen and Male Buffalo			5721	3026	2925
Cows	90,414				
Female Buffalo	1,10,089				
Youngstock			91,858	30,367	40,481
Total Cattle	3,48,902	3,38,074	4,31,746	1,64,557	1,97,963
Ploughs		63,496	67,361	38,584	43,365
Carts		21,389	23,622	17,288	17,466

Adapted from Nilananda Chatterjee (1926) *The Condition of Cattle in India*, p. 296

Driven by Malthusian and Utilitarian concerns, Indian Famine Commission (1901) also went ahead to desire that private cattle camps, run by native philanthropists, should rather be conducted on scientific lines, and those ones that could pass this rational and utilitarian test, "should be liberally supported by Government in every way". However, they were quick to remark that (Table 3.1):

> But generally, these private camps do not discriminate between useful and worthless cattle, the practical result too often being merely to afford *euthanasia* to useless kine. While we regard with respect, the sentiments which inspire endeavours of this kind, we feel bound to point out that they can hardly be considered successful from the economic point of view.[42]

3.6 Cattle Mortality Due to Contagious Diseases

Not much had improved on the front of cattle mortality statistics from contagious diseases till the year 1905–'06, as becomes obvious when one looks at the annual administrative report of the Civil Veterinary Department of India:

> The returns are not only unreliable, but quite misleading. If this required any proof, it is forthcoming in the fact that statistics collected in the limited area of Surma valley by the Tea Association show that the mortality from Rinderpest alone in the Valley during last year, amongst cattle and buffaloes, exceeded 200,000. This is interesting when read with the report of the Bengal Superintendent, which shows that only 12,934 deaths from contagious disease were reported to his office from the whole of the province during the year. The fact too that 26,45,675 hides were exported from Madras alone during the year also gives some idea of the possible mortality from the disease in the country. The executive and supervising staff is so infinitesimal that we are scarcely able to touch the fringe of the work, and the agency employed for the collection of statistics, although the best that can at present be devised, is extremely unsatisfactory. Reports are submitted by non-professional men who are generally incapable of diagnosing disease and consequently it is impossible even to classify the outbreaks reported under their proper headings.[43]

The epizootic Rinderpest hit Bombay Presidency extremely hard in the year 1906–'07, with "every district, with exception of Broach and Ratnagiri having reported outbreaks". There were 50,144 attacks and 25,416 deaths (50%) were reported. Superintendent in his report remarked, "these figures probably considerably under-estimate the actual losses as there is but little doubt that many attacks and deaths have not been reported at all". Superintendent noted that "the means at the disposal of the Bombay Civil Veterinary Department were *totally inadequate to check an epizootic of this description*" (emphasis added). It was realized that the disease could not be eradicated, and the department made endeavours within its capacity "to limit its mortality as far as possible". Various districts boards were asked to employ temporary Veterinary Assistants in order to carry out preventive inoculation and 36,264 animals were inoculated. Surprisingly, Kaira district accounted for only 52 inoculations and the superintendent considered this as very disappointing and remarked that "a man of Veterinary Graduate V. D. Patel's experience, influence and service in the district should have done much more".[44]

Superintendent K. Hewlett had also submitted a detailed report regarding Rinderpest to Government on 17th June 1907.[45]

The annual administrative report of the Civil Veterinary Department for the year 1907–'08 had once again voiced under-confidence regarding the statistics reported on cattle mortality owing to contagious diseases by stating:

> The question of collection of statistics regarding the mortality from contagious diseases amongst cattle is an exceedingly difficult one owing to the fact that there is practically no agency which could be utilized where the officials have not already sufficient work to keep them fully employed. It is practically impossible for the Subordinate Veterinary Department to do the work as areas under their charge are so large and, in many provinces, only a very small proportion of the sanctioned establishment has so far been appointed. Whatever agency is employed is therefore a non-professional one, unable to correctly diagnose the disease which may occur. It has already been noted in previous reports that the figures do not in any way represent the total mortality from contagious diseases and specific instances have been adduced to prove this.[46]

During the year 1907–'08, there were 41,598 cases of attacks of Rinderpest in Bombay Presidency leading to 19,413 deaths. In his annual report, the superintendent noted, "The severity of the outbreak appears to have slightly decreased".

The annual administrative report of the Civil Veterinary Department of India for the year 1908–'09, reported the total number of deaths as 4,01,139 heads of bovine animals; of which 2,38,374 were from Rinderpest; 26,473 from Foot and Mouth disease; 34,124 from Haemorrhagic Septicemia; 10,299 from Black Quarter; 17,790 from anthrax and 74,080 from other contagious diseases.[47]

By the year 1910–'11, the demand for Rinderpest anti-serum was noticed to be increasing with each passing year and it was stated to be far greater than the output that can be obtained at Imperial Bacteriological Laboratory, Muktesar. The annual report of Civil Veterinary Department for 1910–'11 described the year as "a particularly busy one" for Imperial Bacteriological Laboratory, Muktesar and stated that "the manufacture of serum was carried out under extreme pressure both as regards the staff and the supply of bulls".

By 1910–'11, Superintendent of Civil Veterinary Department in Bombay Presidency, Lieut. Col. H. M. Maxwell had outlined that the work of inoculation needed to be taken up with much more energy and it would be great if a Deputy Superintendent post is sanctioned so that this task can be taken up with much more seriousness. To stress upon this point, he stated in his annual report:

> The only way to get a thorough control of an outbreak is for a number of men to attend at once and inoculate every head of cattle in the affected and all cattle herds. To do this, we require more trained men, but what is still of greater importance, the prolonged attendance on the spot of a Superior Officer, who from his thorough knowledge of the language and the people can so represent the position of affairs as to induce all owners to combine and have every animal protected at once…I hope the appointment of a Deputy Superintendent will shortly be sanctioned whose chief duties will be to organize and establish a systematic way of dealing with outbreaks of contagious disease. I wish to urge the deep consideration of this matter as the annual loss from rinderpest alone is very heavy and this is a matter of deep regret when we know that if only outbreaks can be properly dealt with, the disease can be arrested.[48]

This request by Maxwell was acted upon and Mr. J. D. Buxy was appointed as the first Deputy Superintendent of the Civil Veterinary Department of Bombay Presidency.

During 1914–'15, the number of animals attacked by Rinderpest was 16,560 and 6,833 of them died in Bombay Presidency. Superintendent, Major G. K. Walker noted that "Epizootics among cattle and more especially rinderpest and foot and mouth disease may be frequently traced to infection from adjoining Native States where there is little or no veterinary staff available" and observed that "at present restriction of movements of cattle appears to be impracticable".[49]

The year 1915–'16 was once again marked by serious and widespread prevalence of rinderpest in Bombay Presidency and the mortality due to this contagious disease was 13,168; i.e. almost double compared to the previous year (6833). In his report Superintendent, Major G. K. Walker conjectured that "the serious outbreak in the south of the Presidency which commenced in the previous year in the Hangal taluka of Dharwar district was imported from Mysore". It was now increasingly felt that "without legislation and a very large staff it is impossible to prevent further recrudescence or guard against reinfection".

At the first meeting of Veterinary Officers in 1919, it was pointed out by Ware that "cattle Disease Act (1866) had proved valuable in Madras in dealing with cases of cattle disease at fairs and elsewhere, but circumstances frequently prevented it from being put into operation"[50]. The discussion on the theme of 'the necessity for further legislation' at this meeting does suggest that while legislation was being sought, different officers had presented different rationale justifying the need and desirability of the one. They also gave the impression that they were not sure about the practicability of the legislation at that point in time. While Colonel Walker thought that legislation was bound to come, more particularly in connection with enforcing inoculation; Mr. Oliver thought that it might be advisable to obtain legislative powers to prohibit movements and thereby compel owners to agree to inoculation before permitting movements of their animals.[51]

By the year 1920, concerns around epizootics had started to get voiced in official circles in Bombay Presidency. At a meeting of the Provincial Board of Agriculture held at Poona on 9th and 10th June 1920 the theme, 'Prevalence of Epizootic Diseases Amongst Cattle and the Policy to be adopted with regard to them' entered the agenda. The discussion at the meeting started with a note prepared by G. Taylor, Superintendent, Civil Veterinary Department, Bombay. Taylor's note had expressed a tone of disarming frankness that it was "rather late in the day" to pay attention to this debilitating threat:

> To any one directly interested it would seem rather late in the day to question the prevalence of epizootic diseases amongst cattle in this country, and when I suggested the subject of their control as suitable for discussion by the Board of Agriculture, I did not consider that it would be necessary to emphasise the gravity of the position. Until comparatively recently the absence of cut-and-dried statistics, and the fact that the experience of those directly affected remained to a great extent unexpressed, would excuse the failure of many to realise fully the annual loss that is occurring from this cause.[52]

Taylor argued in his note that the mortality statistics published in the annual reports of the Veterinary Department provided a speculative indication of the position, even as he hesitated to accept them as complete. He had appended with his note a statement of the figures on mortality statistics available for the years 1909–'10 to 1918–'19 and considered that to be sufficient to dispel any doubt that the control of contagious diseases required urgent attention.

During the discussion, B. S. Kamat voiced his opinion that "people were not yet ready for legislation either to make the report of diseases compulsory or to make inoculation compulsory". He also posed a question "whether legislation of the nature proposed, which had been undertaken in Madras and Burma, was not a dead letter", to which Taylor replied that "the Madras (legislation) had remained a dead letter until the Veterinary Department in Madras was strong enough to deal with epizootic diseases, but now the Veterinary Department in Madras had been strengthened, the legislation was found very useful".[53] Participating in the discussion Dadubhai Desai proposed that "a veterinary dispensary should be established in each taluka and until that had been done there was no use introducing legislation". Purushottamdas Thakurdas, however voiced the oft repeated complaint that "the quality of the cattle in this (Bombay) Presidency was fast deteriorating, and that Government was to some degree responsible for it". Asking experts' opinion on this view, he expressed his view that such a deterioration "was due to the slaughter of good milk animals in Bombay, which are sold to butchers when they go dry".[54] Purushottamdas Thakurdas in this discussion comes across as quintessentially speaking on behalf of cow-protectionist groups, who kept agitating against the government on the issue of cattle slaughter, and linked its fallout on the milk becoming steadily scarcer and scarcer. At the same time, it is surprising that he failed to get agitated by the prospects of the loss of thousands of agricultural cattle in the countryside from the contagious diseases. However, when one reads Taylor's note, one doesn't fail to grasp that he had half-guessed the response from 'native' colleagues on provincial board. Voicing his frustration that control of contagious cattle diseases by way of inoculation had not agitated public opinion as much as it should have, he wrote:

> Regarding public opinion, it is interesting to observe how quite lately it has been given expression to in one way and another. While the direct loss due to the death of animals from disease still remains unremarked, except in official documents, the stringency of milk supply is exercising the minds of the urban and literate classes with the result that a good deal of attention from time to time has been called to the depletion of herds by slaughter or export. It is strange that attention should alight on factors the good or evil economic influence of which is a debatable, while the enormous dead loss from contagious disease should excite little or no comment.

By the year 1921, the serious loss of bovine stock from contagious diseases and the deleterious impact it might have on the fate of agriculture had not gone unnoticed, even in contemporary writings that talked about the great services rendered by Army Veterinary Corps in the First World War. Hinting at the comparison between the loss resulting in Army animals and cattle belonging to 'civil community' from the deadly contagious diseases, Major General Sir John Moore wrote in his book, *Army Veterinary Service in War*:

> In the section 'Wastage of Animals in War' I showed, as briefly as I could, the serious amount of loss resulting in the Army from dire contagious animal diseases. *This is nothing compared to what occurs in Civil Communities in India, upon which the Army is dependent for its Transport and Food Supply.* In the Civil Veterinary Statistical Returns for the year 1917-'18, 3175 Equines; 2,49,010 Bovines and 22,375 other animals are shown as having died from contagious diseases alone. In the year 1918-'19, 933 Equines; 2,28,414 Bovines and 12,858 other animals are similarly shown, and Reports at the same time make it quite clear that *owing to the weakness of the cadre of the Civil Veterinary Department, and to the inadequacy of the machinery necessary for complete control, full mortality is not recorded.*[55] (emphasis added)

During the year 1920–'21, the number of deaths due to Rinderpest recorded was 5910, as against 6870 in the previous year, even as the deadly contagious disease was reported from all the districts of Bombay Presidency. Talking about the unreliability of these statistics, G. Taylor, Superintendent of Civil Veterinary Department of Bombay Presidency stated:

> These figures cannot be relied on as showing that the year has been a comparatively healthy one, so far as rinderpest is concerned. The facts are the reverse, and the disease has been particularly widespread. Last year there were only 452 outbreaks of rinderpest reported while this year the number was 793. The mortality from rinderpest as shown by the figures of the epizootic disease reports submitted by the staff when they investigate an outbreak is 9,479 which is nearly double that of the village officers' reports on which we depend for our statistics, notwithstanding the fact that the former are usually based on a single visit whereas the latter are supposed to be weekly returns of the mortality throughout the duration of the outbreak.[56]

With a view to obtain some information on the rinderpest outbreaks that would give somewhat accurate information of the losses, special investigations were carried out by Deputy Superintendent, Khan Saheb Buxy in Surat district and similarly Veterinary Inspector Nagarsheth had carried out special investigations in Nasik district.

In Bombay Presidency, deaths from Rinderpest suddenly surged upto 10,593 in the year 1921–'22, when the attacks were reported from all the districts except Belgaum and Bijapur. The next year, 4767 deaths were recorded, since this contagious disease was not as widespread as in previous years, but it only made sporadic appearance in all districts except East Khandesh and Belgaum. During the year 1923–'24, Rinderpest got reported from all districts of Bombay Presidency except Panch Mahals and Dharwar and number of deaths recorded from it was 5166. Superintendent noted in his annual report that "it was more widely prevalent during the year owing to the scarcity of the fodder, on account of which there were free movements of the affected herds of cattle of graziers, in search of grazing".

The question of 'Methods of Obtaining and Recording Statistical Information Regarding the Incidence of Epizootic Diseases in India and Suggestions for Improvement' again came up for the discussion at the second meeting of Veterinary Officers in India held at Calcutta in 1923. Introducing the subject, the chairman stated:

> In perusing the mortality figures in our annual reports, I feel that inaccuracy is rampant, but we are in the hands of informants entirely. The system of collecting statistical information vary in different provinces. I shall be interested if any representative regards his local system as satisfactory. Statistics relating to diseases should be checked by professional men, if possible, *but our numbers are too few*. (emphasis added)

The meeting passed the following resolution after the discussion over the subject:

> In view of the administrative and professional importance of obtaining reliable statistics in regard to the mortality incidence in animal disease, a more thorough system of notification than the present very unsatisfactory system of reporting should be immediately investigated and introduced. The system approved, should, as far as possible, be uniform in its application throughout India.

The year 1924–'25 was very severe and Rinderpest attacks occurred with the exception of Panch Mahals district, in the whole of the Bombay Presidency. 1950 villages were reported to have suffered from Rinderpest and 36,534 cattle died out of 77,478 attacks. Out of 1950 villages attacked by Rinderpest, 975 outbreak villages accounted for preventive inoculation and a total of 1,17,358 animals were protected. Superintendent remarked in his annual report that, "this outbreak has shown the necessity for (1) a larger mobile force of Veterinary Assistant Surgeons than we have at present and (2) the necessity for some form of legislation to deal with cattle disease". Making a comment on lack of needed human resources in department, he stated, "as most of men employed are in charge of stationary dispensaries, it is difficult to transfer them from their own districts for special work, and the only extra hands available are reserves" and he felt that "the number of these (reserves) is too small for the purpose". It was corollary to this realization that "proposals for the more rapid expansion of the Department and for introducing legislation for the control of contagious diseases of cattle, in connection with the recommendation of the cattle breeding committee were submitted to the Government".[57] However in the next year's annual report, the superintendent had to note once again: "At the request of the Government, proposals for the introduction of legislation for the control of contagious diseases were submitted for consideration". Superintendent stressed upon the point that "until some such action is taken, very little can be done to prevent these diseases from spreading and causing considerable loss annually".[58]

While around 1910–'11, the veterinary officials in India had felt that they had found a way to stamp out Rinderpest by engaging with the propaganda around the need for inoculation, by mid-1920s they increasingly felt and stressed the point in official correspondence that such a measure without being accompanied by legislative powers that makes reporting of contagious disease and inoculation compulsory would have the potential to wash out the gains. The subject of legislative powers to address the question of contagious diseases amongst cattle was also on the agenda of the First meeting of Veterinary Officers in India, which was held at Lahore in March 1919, "but the Government of India doubted whether with the existing staff any practical advantages would be gained by enhancing the legal powers of enforcing preventive and curative measures against cattle disease", reported Col G. K. Walker, who presided over the Second Meeting of Veterinary Officers in India, that took place in Calcutta from 26th February to 2nd March 1923, in his opening address. While calling attention of his colleagues on the theme 'the desirability of introducing further legislative measures for the control of animal diseases', Col G. K Walker added that "the position has not altered materially since then (i.e. since the time Government of India pronounced that response to the discussion on the subject at the first meeting of veterinary officers in India)".[59]

Similar thoughts about the necessity of a properly constituted Contagious Diseases Animals Act were articulated by Major General Sir John Moore, the then Director of Army Veterinary Service in India in his book titled *Army Veterinary Service in War* in 1921. While talking about the grave contagious diseases prevalent in India, Moore (1921: 80–81) stated:

> The dire diseases to which I refer are Rinderpest (cattle plague) and Foot and Mouth disease in Cattle, Surra in Camels and Anthrax in various animals. *The history of these diseases and the dead loss sustained not only in army, but in Civil Communities has been and still is appalling, and the machinery of control is most inadequate and imperfect in operation.* One would have thought that with the mass of contagious animal disease of all kinds in India, worse by far than any other country that I am acquainted with, one of the first considerations would have been a properly constituted Contagious Diseases Animals' Act to suit the necessities of the country, *but such is not the case, and animal life continues to be wasted.* It is not for me to criticize the policy, but I am all out to assist in the grading up of animal efficiency and in the reduction of the wanton, unnecessary animal wastage and suffering[60].

The enormity of loss from contagious diseases can be gauged by the statistics John Moore cites drawing from Civil Veterinary Statistical Returns for the years 1917–18 and 1918–19, respectively. Talking about the inadequacy of the machinery necessary for the complete control, Moore (1921: 94) stated:

> In the Civil Veterinary Statistical Returns for year 1917-18, 3175 Equines; 2,49,010 Bovines and 22,375 other animals are shown as having died from Contagious Disease alone. In the year 1918-19, 933 Equines; 2,28,414 Bovines and 12,858 others are similarly shown, and Reports at the same time make it quite certain that *owing to the weakness of the cadre of the Civil Veterinary Department, and the inadequacy of the machinery necessary for complete control, full mortality is not recorded.*[61]

John Moore believed that while "India has made very rapid strides in Agriculture during the past ten years (i.e. 1911–1921), but the Animal Husbandry side of Agriculture has not kept pace with the growth of crops. *Newspapers teem with reports and literature regarding wheat and cotton, but not a single word ever appears about (live)stock. In my opinion, it is really a very sad situation, and the danger of it is that there may come a day when the country wakes up to the fact that through heavy mortality, there is a grave shortage of animals necessary for the life of the country*". (emphasis added).

John Moore was not merely being alarmist, but he was concerned and increasingly felt that there needed to be a coordinated effort between Civil Veterinary and Agriculture Departments and the Veterinary Service of the Army to address this issue. The sentences he writes following the above premonition indicate a zeal to carve out such a cooperation:

> The Army through its Educational policy can be made a very powerful factor for good in the situation described. By a combination of Civil Veterinary and Agricultural Departments and the Veterinary Service of the Army, it ought not to be difficult to draw out some plan of campaign whereby men, before taking their discharge and returning to civil life, in rural districts, could receive a little up-to-date instruction in animal husbandry, dairying, farm economies, etc... It is to be hoped in the interest of stock that such measures can be introduced soon. The Army with its Veterinary Service on a self-contained basis, complete with its own trained personnel, and with its Veterinary Hospitals properly organized and conducted, will have every facility for carrying out its desired programme in conjunction with the Civil Authorities.

During the year 1925–'26, Rinderpest was reported from 1382 villages, attacked 52,465 animals and caused 23,465 deaths in Bombay Presidency. Similarly, Rinderpest was the cause of heavy casualties amongst cattle amounting to 14,892 deaths in Bombay Presidency, during the year 1929–'30. The year 1930–'31 was also an unhealthy one for livestock in Bombay Presidency, with Rinderpest being very widespread in Dharwar, Belgaum and Kanara districts. There were 57,603 attacks and 28,187 deaths and many casualties occurred in Ahmedabad district. E. S. Farbrother, Superintendent noted in his annual report emphasizing on the necessity of timely reporting of contagious diseases:

> Rinderpest continued to spread in the Presidency and to cause heavy losses. In previous reports, I have stated that reporting of outbreaks had considerably improved, but in Ahmedabad district this year, disease gained a very strong foothold, before it was reported properly, and the mortality was consequently very heavy.

During the year 1932–'33, Bombay Presidency reported 58,225 attacks and 29,170 deaths due to Rinderpest. During the year 1936–'37, Bombay Presidency reported a more widespread prevalence of Rinderpest, which swept away 20,725 animals.

At the second meeting of Animal Husbandry Wing of Board of Agriculture and Animal Husbandry in India, there was a discussion that aimed "to obtain the consensus of opinion as to the best method of employing goat virus whether as a prophylactic or purely as an outbreak stopping method". At this discussion, F. Ware took a position that was typically rooted in the notion by which cattle were being viewed as 'useful' and 'useless' and veterinary administration was feeling compelled to only preserve those cattle that were considered valuable by them. While there were other officers, such as Mr. Mitchell, who were not in favour of using vaccine as an outbreak stopper only, F. Ware "suggested that only the best cattle should be vaccinated and that inferior animals should be allowed to take their chance seeing that *there was a great surplus of cattle in the country and scarcity of fodder*". (emphasis added) Mr. Mitchel had used "an analogy of small-pox immunization in human beings to illustrate his point that protection should be extended as widely as possible". While the advancement in veterinary research had made it possible to provide permanent immunity from Rinderpest by spending less than a farthing—while earlier using the serum simultaneous or serum alone method used to cost about 30 rupees—F. Ware proposed that "by charging a small fee for vaccination, owners would be forced to decide which of their cattle were worth paying for and which might be left unvaccinated".

However, such a Malthusian thinking voiced by F. Ware was met with differences being voiced by his colleagues. Joining his voice with Mr. Mitchel, "Mr. Garewal considered that now that a successful vaccine had been evolved after years of experimentation, the widest possible use should be made of it in an anti-rinderpest work". Mr. Garewal forcefully argued "whether it would be proper to allow inferior cattle, which might represent the only resources of poor peasants, to die; particularly when an efficient protective agency had at last been obtained". Mr. Riazul Hassan also felt that "with the means of protection now available, *no cattle should be allowed to die of Rinderpest*". (emphasis added) The proceedings also notes that "Mr. Parr saw great danger in Mr. Ware's policy and questioned whether so-called scrub cattle were always surplus". Mr. Parr had urged his colleagues to exercise extreme caution, because of the economic factors involved in such a policy. While Mr. Kerr voiced an opinion similar to the one articulated by F. Ware and argued that "certainly in Bengal, a large proportion of cattle were surplus to requirement, and not worth preserving", Mr. Shirlaw reminded that "this policy also had a moral bearing in that the question of medical ethics was undoubtedly involved".

The stand taken on this discussion by Colonel Arthur Olver was very remarkable since he asserted that "Mr. Ware's suggestion that inferior cattle should be allowed to die could not be accepted by meeting" especially "as it became more generally known that cattle could safely and cheaply be protected against death from Rinderpest". Olver felt that quite contrary to how Mr. Ware thought about the issue, the assurance of cheaper mode of providing prophylactic inoculation against Rinderpest might perhaps induce in stock owners a different rationality where they "would realise that it was no longer necessary to maintain surplus cattle as an insurance against periodical outbreaks of Rinderpest".

Amongst all the bovine diseases, Rinderpest continued to be the fatal disease as of 1938–'39. The annual report, 'Agriculture and Animal Husbandry in India' noted that "rinderpest still claimed more than half of the total loss"[62]. The share of mortality due to rinderpest in Assam, Orissa and the United Provinces ranged from as high as 63 to 70 percent of total mortalities, while in Coorg, Bombay, Sind, Ajmer-Merwara and Madras the proportion of mortality from rinderpest was 30 percent or lower.[63]

Outbreaks of diseases had also continued to mar the prospects of cattle shows and exhibitions in the year 1938–'39, when the second All India Cattle Show was held in Delhi. The annual official report, 'Agriculture and Animal Husbandry in India' noted that "Out of 27 breeds scheduled in the prospectus, five were not represented owing to certain unavoidable circumstances". The report suggested that one of these unavoidable circumstances was "outbreaks of foot and mouth disease".[64]

The next chapter talks about the discussions amongst colonial experts on the question of cattle breeding policies.

Notes

1. See for details, Samanta, Samiparna (2016) 'Cattle, Cruelty, Cow Doctors: Examining Animal Health in Rural Bengal, 1850–1920', in Deepak Kumar and Bipasha Raha (eds.) *Tilling the Land: Agricultural Knowledge and Practices in India*, Primus Press, New Delhi. Also see, chapter 3, titled 'The Politics of Care' in Samanta, Samiparna (2021) *Meat, Mercy and Morality: Animals and Humanitarianism in Colonial Bengal (1850–1920)*, Oxford University Press, New Delhi, pp. 62–132.
2. See, Queripel, A. E. (1879) 'Cattle plague in India' *The Veterinary Journal and Annals of Comparative Pathology*, Vol IX, Issue 11, November 1879, pp. 310–316.
3. The opening speech by Colonel Walker while introducing the subject, 'The necessity for Reorganisation of the Indian Civil Veterinary Department', at the First Meeting of Veterinary Officers held at Lahore on the 24th of March 1919 and following days.
4. The first Annual Administrative Report of the Civil Veterinary Department in India for the official year 1892–93 had presented a short history of each of the several Veterinary Colleges and Schools operating in different parts of India at that point of time.
5. See, Voelcker, John Augustus (1893) *Report on the Improvement of Agriculture in India*, p. 212.
6. The first Annual Administrative Report of the Civil Veterinary Department in India for the official year 1892–93, p. 40.
7. Col Trevor had addressed several Darbars of Rajputana in order to enroll their financial support and by 1891 he was promised Rs. 6504 as annual subscriptions in addition to being hopeful that "there would be no difficulty in raising the sum of Rs. 10,000 for the construction of college building at Ajmere (sic) if required". For the outline of Ajmere Veterinary School see, Letter from J. H. B. Hallen to the Secretary to the Government of India, No. 37, Revenue and Agriculture Department, dated 1st February 1891.
8. Smith, Frederick (1927) *A History of Royal Army Veterinary Corps* (1796–1919), Bailier, Tindall & Cox, London, p. 206. Smith recalls this period when Horse Breeding Operations of the Indian Studs, which had then been in existence for 80 years (i.e. since 1793), were handed over to a news appointed commission that had J. H. B. Hallen as one of the commissioners in following words: "In 1902, the Civil Veterinary Department in India received a serious setback relative to Horse Breeding establishments. *The very words, 'Horse Breeding in India' stink in the nostrils of those who know the story. Its history requires a volume to itself, which the Author hopes one day to furnish.* In 1873, the old Indian Studs had been swept away after 80 years of organization, reorganization and disorganization, and a scheme devised by Lt Colonel Hallen of the Indian Veterinary Service was instituted. *During the many years, he supervised this work there were no complaints so far as I am aware, but after his retirement for age in 1894, opportunity was taken to condemn his system as having "completely failed to answer the expectations, it was intended to fulfil" and reorganization was demanded.* A Commission sat in 1900–01; they attributed the shrinkage in remount supply to the officers of the Civil Veterinary Department, who, it was stated, had proved unsuitable for the work. *"Their training does not specially fit them for it; no practical knowledge of breeding is gained by them during their veterinary training, and a few of them have more than limited knowledge of Native Languages*". The reorganization recommended by the Commission was a partial

reversion to the old Stud system, which had so lamentably failed, viz. the removal of horse breeding from civil veterinary to military remount control, the Civil Veterinary Department being left to concentrate on education and the study of the diseases of stock. Behind this desire for reorganization was the feeling expressed in the statement, freely repeated by the Commission, that "friction has always existed between the Remount and Civil Veterinary Departments," and therein doubtless lay the reason for the outcry for reform.

9. For a short sketch of the history of this institution, now renamed as Mumbai Veterinary College, please see: https://mvc.ac.in/about-us/bombay-veterinary-college/.
10. The first Annual Administrative Report of the Civil Veterinary Department in India for the official year 1892–93, p. 43. For a copy of this correspondence, see Appendix V, p. 114.
11. Cited in, The first Annual Administrative Report of the Civil Veterinary Department in India for the official year 1892–93, p. 43.
12. Smith, Frederick (1927) *A History of Royal Army Veterinary Corps* (1796–1919), Bailier, Tindall & Cox, London.
13. 'The Indian Rinderpest', *The Calcutta Review*, Vol 55, No 109, Pp. 169–186.
14. Annual administrative reports of the Bombay Civil Veterinary Department filed by J. K. Grainger, Superintendent, Horse Breeding Operations, Bombay Presidency during his tenure (1st April 1887–19th May 1892) only shows details on horse breeding operations and participations in Horse Shows. There is not a single word on cattle diseases to be found in any of these reports! In 1892–93, for the first time we come across the sign that the officer in charge of Civil Veterinary Department in Bombay attempted something else in addition to his primary work on overseeing Horse and Mule Breeding Operations, when we find a mention on his experiments in fodder crops and sheep breeding at Ahmednagar, albeit instituted at his own expense. See, Annual Administrative Report on Civil Veterinary Department for the year 1892–93, pp. 12–13.
15. Editorial Remarks in *The Veterinary Record*, Vol IV, No 185 (January 23, 1892), p. 385. Curiously, these editorial comments were followed by the weekly journal carrying the text of the Office Memorandum signed by W. B. Walters of Army Veterinary Department verbatim. See, *The Veterinary Record*, Vol IV, No 186 (January 30, 1892), p. 401. Not only was the text of the Office Memorandum appeared, but right beneath it, the weekly journal carried a note of clarification and some further details of the scheme.
16. Editorial in Madras Mail dated 20th February 1892, republished in *The Veterinary Record*, Vol IV, No 194, (26th March 1892), Pp. 517–518.
17. See, Mishra, Saurabh (2011) 'Beasts, Murrains and the British Raj: Reassessing Colonial Medicine in India from the Veterinary Perspective, 1860–1900'; *Bulletin of History of Medicine*, Vol 85, No 4, Winter 2011, pp. 587–619.
18. In 1919, out of 42 appointments/ positions, there were 12 vacancies, and 7 officers were engaged in military duty, thus 19 vacant posts. Mackena, J (1919) 'Opening Lecture to the First Meeting of Veterinary Officers in India', *Proceedings of the First Meeting of Veterinary Officers in India*, Superintendent of Government Printing, Calcutta. p. 05 and p. 20.
19. Ibid. p. 06.
20. Ibid. p. 06.
21. Review comments by C. S. Curtis on the Annual Administrative Report of the Civil Veterinary Department of Bombay Presidency for the year 1894–95.
22. Ibid. p. 05.
23. See, Voelcker, John Augustus (1893) *Report on the Improvement of Agriculture in India*, p. 212–213.
24. Annual Administrative Report of the Civil Veterinary Department of India, 1893–94, Government of India Press, Calcutta, p. 84 & 85.
25. Annual Administrative Report of the Civil Veterinary Department of India, 1895–96, Government of India Press, Calcutta, p. 145.
26. Quoted in Annual Administrative Report of the Civil Veterinary Department of India, 1896–97, Government of India Press, Calcutta, p. 157. Also see Annual Report of the Civil Veterinary Department, Bengal for the year 1896-'97, p. 13.

27. Annual Administrative Report of the Civil Veterinary Department of India, 1897–98, Government of India Press, Calcutta, p. 159.
28. Mishra, Saurabh (2015) 'Cattle, Famines and the Colonial State' in *Beastly Encounter of the Raj: Livelihoods, Livestock and Veterinary Health in India*, Manchester University Press, p. 80.
29. Famine Commission's report in 1880 remarked, "Of hides, raw or tanned, some eight millions are exported every year, valued at about three million sterling. The trade received an impulse in 1877–'78, in which year about one and a half million hides above the usual number were exported, an increase probably due to the deaths of cattle in consequence of drought of 1876.".
30. Report of the Indian Famine Commission (1898), p. 188.
31. Quoted in The Annual Administrative Report of the Civil Veterinary Department of India for the year 1899–1900, p. 69.
32. The Report of the Indian Famine Commission 1901, p. 72.
33. The annual report on Cultural, Dairying and Breeding Operations conducted on Experimental Farms and Gardens in Bombay Presidency, 1901–'02, p. 23.
34. The Annual Administrative Report of the Civil Veterinary Department in Bombay Presidency for the year 1901–'02, p. 09.
35. Quoted in Annual Administrative Report of the Civil Veterinary Department of India, 1899–1900, Government of India Press, Calcutta, p. 69.
36. Compared to these figures, the Pinjrapol at Ahmedabad managed by the Mahajan accommodated in all 21,543 cattle and 202 horse and "being run entirely on religious lines, no animal, however emaciated or vulnerable, was refused admission and food". However, mortality rate at Pinjrapol was high and 16,712 animals died.
37. Annual Report of the Civil Veterinary Department of Bombay Presidency, 1899–1900, p. 18.
38. See Appendix 54, in Report on the Famine in the Bombay Presidency (1899-–1901), Vol II, Appendices, pp. 169–170.
39. Report of the Indian Famine Commission (1901), p. 75.
40. Report on the Famine in the Bombay Presidency 1899–1901, Vol II, Appendices. Review of the Cattle Camp Scheme by The Commissioner, Northern Division. p. 164–165.
41. Report of the Indian Famine Commission (1901), p. 75. This recommendation was accepted wholeheartedly and entered as a guideline prescribed in Famine Code. For example, see Paragraph 203 titled 'Cattle camps' in Famine Code for United Provinces (1912: 37): "When a fodder famine is severe a cattle camp may, with the previous sanction of the Local Government, be instituted in each district under veterinary supervision. Only selected cows and a few selected bulls shall be accommodated in the camps, the chief object of which is to preserve valuable breeds. The collector may purchase the cattle admitted to the camp, and at the end of the famine re-sell them.".
42. Report of the Indian Famine Commission (1901), p. 75.
43. Annual Administrative Report of the Civil Veterinary Department of India, 1905–06, Government of India Press, Calcutta, pp. 05–06.
44. Annual Administrative Report of the Civil Veterinary Department of Bombay Presidency, 1906–'07, p. 30.
45. See, Report No A-3256, dated 17th June 1907. The mention of this report is found in Review of the Annual Administrative Report of the Civil Veterinary Department of Bombay Presidency, by P. J. Mead, Acting Director, Department of Agriculture Bombay Presidency dated 15th August 1907.
46. Annual Administrative Report of the Civil Veterinary Department of India, 1907– '08, Government of India Press, Calcutta, p. 08.
47. Annual Administrative Report of the Civil Veterinary Department of India, 1908– '09, Government of India Press, Calcutta, p. 09.
48. Annual Administrative Report on the Civil Veterinary Department of Bombay Presidency for the year 1910–11.
49. Annual Administrative Report on the Civil Veterinary Department of Bombay Presidency for the year 1914–'15, p. 20.

50. The history of cattle diseases and the veterinary medicine and education in Madras throws up several ups and downs. At the time of Voelcker's visit to India (1890), Government of Madras Presidency had "abolished the cattle disease branch of their Agriculture Department", a decision that Voelcker had characterized as exhibiting "retrograde character". Voelcker's inquiries found that "the veterinary staff was insufficient and inefficient, and that the cultivators offered opposition to the action of veterinary officers", which had in turn compelled the government to give up "for the time, all attempts to cope with epidemics". See, Voelcker, John Augustus (1893) *Report on the Improvement of Agriculture in India*, p. 212.
51. Proceedings of the first meeting of Veterinary officers in India held at pp. 33–34.
52. Proceedings of the Bombay Provincial Board of Agriculture held at Poona on 8th and 9th June 1920, p. 17–18.
53. Madras had passed a provincial Cattle Disease Act in 1866.
54. Nilanda Chatterjee's 1926 book, *The Condition of Cattle in India*, which was published by a Calcutta based cow-protection group cites statistics on animals slaughtered at Bandra and Kurla municipal slaughter houses from April 1919 to March 1924, based on annual returns. It argues that during this five year period; 2,29,257 cows; 73,636 bullocks and 90,804 buffaloes were slaughtered in Bombay. It also estimates that 97,583 calves and young buffaloes were tortured to death in several milch stables that dotted Bombay municipality.
55. Moore, John (1921) *Army Veterinary Service in War*, H&W Brown, London, p. 94.
56. Report of the Civil Veterinary Department of Bombay Presidency for the year 1920–'21, p. 16.
57. Report of the Civil Veterinary Department of Bombay Presidency for the year 1924–'25.
58. Report of the Civil Veterinary Department of Bombay Presidency for the year 1925–'26, p. 26.
59. See, Proceedings of the Second Meeting of Veterinary Officers in India held at Calcutta from 26th February to 2nd March 1923, p. 08.
60. Moore, John (1921) *Army Veterinary Service in War*, H. & W. Brown, London, pp. 80–81.
61. Moore, John (1921) *Army Veterinary Service in War*, H. & W. Brown, London, p. 94.
62. 1,17,400 deaths were reported due to Rinderpest during the year 1938–'39 as against 89,700 deaths reported due to Rinderpest in the previous year. Agriculture and Animal Husbandry in India for the year 1938–'39, p. 182 and p. 218.
63. Ibid, p. 182.
64. The report stated that the breeds which were not exhibited were: Ongole, Hallikar and Krishna Valley breeds from Madras, Mehsana buffaloes from Baroda and Nagpuri buffaloes from Central Provinces. When the report makes an assertive claim, almost sounding like self-patting that "It is noteworthy that among such a large gathering of animals from all over India, there was no case of any contagious disease and not a single casualty", one does get a feeling that the veterinary officers in the charge of cattle show still carried the dark memories of an outbreak of cattle plague (rinderpest) during Calcutta Cattle Exhibition in the year 1864. Agriculture and Animal Husbandry in India for the year 1938–'39, p. 175.

Chapter 4
Breeding Bovine Animals for Better Productivity: High Modernist Mission

As argued in the introduction, within the history of cattle development in India, there is relatively less engagement with the question of breeding policies related to agricultural stock and milch stock and what transformations were witnessed during the late colonial period. Recent history writings that look at the cattle in India from the lens of veterinary medicine and disease control argues that even after the creation of Civil Veterinary Departments in British India, there was very little attention paid to epizootics amongst 'public cattle' as opposed to the attention that animals important for military (horses and elephants) received.[1] However, one also needs to remember that certain cattle breeds such as Amritmahal of Mysore state also possessed certain traits that proved enviable in military campaigns. This breed was greatly patronized by Mysore Maharaja and Tipu Sultan had also carved out 'protected grasslands' (*kaaval*) to encourage the breeding of Amritmahal bulls and bullocks.[2]

4.1 Tracing the History of Breeding Operations at the Oldest Hissar Breeding Farm

Giving a cursory history of Hissar Farm, J.W.A. Morgan states that "this farm was established in 1809 for camel breeding and the breeding of cattle and horses was added in 1815".[3] Morgan was writing about a breeding farm that had been in existence for around seven decades and within this period it would have seen many ups and downs. However, Morgan's narrative can trace only some faint outlines when he states, "In 1853, it was decided to restrict operations to breeding bullocks for artillery and ordinance purposes, the management being transferred from the Commissariat to the Stud Department. In 1874, at the instance of the Stud Commission, it was again transferred to Commissariat Department and then since 1898, the Civil Veterinary Department had held the charge of the farm". Morgan had also expressed his opinion and thereby stressed upon the need to lay down well-defined cattle breeding

H. Upadhyaya, *Critical Insights on Colonial Modes of Seeing Cattle in India (1850–1980)*, Asia in Transition 27, https://doi.org/10.1007/978-981-97-1560-2_4

policy and ensuring the continuity to achieve the success in what in those years was definitely a long-range work that must be accompanied by a long-term vision and well-articulated strategies. He expressed his criticism of the oscillating policy in these words: "If a definite policy had been formed from the beginning and adhered to, the Farm would have been of much greater advantage to the country than it has been. The breeding of improved stock is a matter of years, in which it is difficult to rectify any error: therefore, definite aims should be kept prominently in view from the start. The objectives of the Hissar Farm were varied at comparatively short intervals up to recent years, so that there was no time to achieve real success in any".

Voelcker had observed in his *Report on the Improvement of Indian Agriculture* (1893: 200) that in the matter of the distribution of stud bull or working bullock, "the system adopted at Hissar, by which the cultivators can go to the Farm and choose exactly what suits their requirement, is decidedly the best one and should be encouraged." While Voelcker didn't cite any other Government cattle farm distributing the bulls to villages, except Hissar Cattle Farm in Punjab, he had advised that "where Brahmani bulls are extinct, or where good country bulls do not exist, then Government can do much good by distribution of good stud bulls" and had added that even if "the results to Government cannot be a directly paying one, it is one which should be undertaken in the interest of the people as agricultural classes".[4] Actually, with disarming frankness, Voelcker exposes the material and financial—rather than welfare—concerns exhibited by the late colonial Government in British India, when he states, "A short time after my visit (to Bhadgaon Cattle Farm in Bombay Presidency), in October 1890, 27 young bulls, were sold at an average of Rs 40 each for breeding purposes" and "this sale alone, enabled the Farm to realize Rs 1080, where as in the Farm report these same animals had been valued at Rs 650 only".

Voelcker (1893: 204) had also held up a warning lantern to the then prevalent administrative policy of appointing the Superintendents to cattle breeding farms under the Military Department for a tenure of five years only and reminded that evils of this system were also observed at Military Grass Farms and *rukhs*. He recommended that "the men who are to hold in future the position of Superintendents should be those who have shown some attitude for the work, and they should receive beforehand the practical training, under the Superintendent, for which the Farm provides the opportunity". He also added that "It would seem desirable to attach an officer of the Commissariat Department from time to time to the Farm, in view of him acquiring, under the teaching of the Superintendent, that knowledge of cattle and of farming operations, generally which all Commissariat officers should more or less possess".

4.2 Perennial Irrigation and Its Deleterious Effect on Cattle Breeding Work in Punjab

In 1910, when Department of Agriculture published *Cattle and Dairying in Punjab*, it was noted while "the great canal colonies have been brought into existence, irrigation from older canals has expanded" there was a corresponding increase in the demand of draught bullocks in last twenty years. It also noticed that even as this demand for draught bullocks was putting a strain on "the unirrigated tracts, which were the chief sources of the cattle supply and were being heavily drawn upon to meet the calls of expanding cultivation", the province was faced by a period of ten years' famine and scarcity. Talking about how the arrival of perennial irrigation had a deleterious effect on the cattle breeding and how this has led to an increasing reliance on the tendency on the part of Punjab zamindar to rely on buying cattle from other tracts:

> As improved communication has gone hand in hand with expanding irrigation, the cultivator finds it easier to buy his cattle than to breed a type suitable to his requirements. The local type is replaced by imported animals, and the distinctive breed disappears. Thus, Captain Pease found that very few pure specimens existed of the cattle formerly characteristic of the Nardak and Bangar tracts in Karnal district… The local breeds gradually disappeared, and their place is now taken by importation from the Hariana tracts of Rohtak and Hissar. Major Walker in his recent survey of Amritsar District states that the Manjha breed described by Captain Pease is now almost extinct as a separate type. Similar forces are at work in the case of the Malwa breed. In the Bar tracts before the advent of canal, little irrigation was done, and the Bar herds though famous for their milch cattle, never produced draught cattle of any importance.[5]

In 1907, in an article published in *The Agricultural Journal of India*, J. Mollison and L. French had argued that if special precautions are not taken to maintain the purity of Montegomary breed, "the breeding of Montegomary cattle is likely to suffer". Qualifying their opinion, they stated that "the extraordinary prosperity across Chenab Colony across the Ravi, has diverted the attention of the Nomadic "Bar" tribe, to the profits derivable from agriculture when assisted by canal irrigation".[6] They had also underlined that since "the whole of the "Bar" tracts of the Montegomary district are destined within next few years to receive irrigation from the projected Lower Bari Doab Canal", the question of how to encourage the nomadic tribe to continue with professional cattle breeding had acquired urgent value.

Responding to a perception amongst native peasants that the quality of cattle breed had deteriorated due to colonial policy that restricted the access to forest and the arrival of perennial irrigation, W. Smith, Assistant Director of Military Dairy Farms, Southern Circle, stated in a paper that he read at the Sixth Indian Science Congress in January 1919:

> It is my opinion, based on thirteen years' close contact with cattle breeding in all parts of this country, the strength of the ox is decreasing in India, and that the country today does not breed the same quality of milk and draught stock, it produced twelve to fourteen years ago. This is due to a variety of causes, but principally, I believe, *to the conserving of many of the old cattle-breeding jungles as forest reserves, and the spread of canal irrigation system, thereby converting what were formerly cattle-breeding jungles into grain-producing areas.*

> If this be so, and I think the fact is generally admitted by those who have studied the subject, then there is only one remedy, which every other country has had to adopt when it met a like problem, viz. the cattle must be bred on the cultivated lands. This can only be done and will only be done when it is economically sound for the cultivator to breed and rear cattle, and it is in order to make the breeding of cattle a payable proposition that the dairy industry must be developed.

Several years later in 1927, W. Smith, who had by now been appointed as Imperial Expert of Dairying, voiced a similar sentiment in a 'Note on Cattle Breeding' that he had prepared for a meeting of the Agriculture Committee in Poona. He once again stressed that "despite all the efforts made in the direction of cattle improvement and scientific cattle breeding, it had become impossible in the open market today to procure in quantity, no matter what the price may be, as good draught bullocks and milch cows as were obtainable 16 years ago". W. Smith felt that the quality of milch cattle available in India, including the Punjab, but excluding Sindh, had become much worse than those available 16 years ago, when he had just arrived in India to work with Military Dairy Farms. Detailing probable reasons for decline in the quality of cattle, Smith underlined that the root cause was, "want of knowledge, accentuated by many circumstances, such as the spread of irrigation canals and conservation of forest lands with consequent diminution of grazing areas, the increased facilities for transport and consequent mixing of breeds or types, increased prices for foodstuffs, and the erroneous idea that the development of dairying or milk production would injure the draught quality of working bullocks".

N. C. Wright had echoed this opinion in his *Report on the Development of Cattle and Dairy Industry in India* (1937: 60–61) by stating:

> At present there is a very general impression that *the introduction of irrigation rapidly leads to the deterioration and even to the virtual extermination of good breeds of cattle.* This, for example, is true of Sindh and certain tracts of United provinces. *If full advantage is to be derived from irrigation, I am convinced that farming in irrigated area will have to be modified to allow the inclusion of mixed farming system in which both crop cultivation and animal husbandry play their part.* (emphasis added)

By the year 1937, the idea of breeding cattle for yielding more milk in addition to giving a good working type cattle had found widespread purchase. Thus, in his opening speech to the second meeting of Animal Husbandry Wing of Board of Agriculture and Animal Husbandry in India, the Governor of Madras stated:

> In providing for the production of working cattle needed for ordinary agricultural operations the necessity for *at the same time producing more milk and of reducing the large number of unprofitable cattle maintained at present should, therefore, not be overlooked.* In Madras, a very good example is provided by the famous Ongole Breed of how, with suitable feeding and management, these requirement can to a large extent be met by the rearing of high grade cattle under semi-stall fed conditions, and *it is of great interest that breeders of such an essentially working type cattle as the Kangayam and the Amrit Mahal, in the breeding of which the attention has in the past been almost exclusively confined to working capacity are now paying more attention to milk production.* (emphasis added)

4.3 Interventions in Cattle Breeding in Bombay Presidency

Till the year 1899–1900, the Civil Veterinary Department in Bombay Presidency had not devoted any attention on the subject of cattle breeding. We find the mention of a consultation that Superintendent had with collectors of 18 districts on this subject in the annual administrative report of the Civil Veterinary Department of India for the year 1899–1900. These collectors were asked to assist, through the medium of the Local Boards, to try and improve the breeds of cattle in Bombay Presidency. Of these 18 collectors, collectors from two districts—namely Surat and Dharwar—had simply said that they would consider the matter, and collectors from Sholapur and Poona showed their willingness to assist in any scheme on this subject. All the other 14 collectors were "*either unable or unwilling to assist in any measure to improve the breed of cattle*" (emphasis added). Morgan speculated that such a response from collectors arose from the peculiar situation that existed in that year of great famine, after two or three seasons of scarcity. He considered that "the present is a particularly bad time to bring forward any measures requiring money, as the finances of most of the local boards show a large deficit owing to plague and famine".[7]

The superintendent had also observed about the regions in Bombay Presidency, such as Krishna valley, Northern Guzerat and the Gir hills of Kathiawar, where the cattle owners were "quite alive to the necessity of keeping their breeds pure" and stated:

> Few finer cattle are to be found in India that exist in the districts named above, and there is no doubt that breeders in these particular tracts are men who thoroughly understand their business, and the greatest care is taken to use only bulls of superior class. The bull calves are castrated when young and sold as yearlings or 2-years-olds to occupants of the land in Kaira, Ahmednagar and Baroda, and on the breeding grounds nothing but a few of the highest class bulls are allowed to graze with the calves and heifers.[8]

Similarly, notwithstanding condescending attitude by certain colonial experts, who thought native Indians didn't know the art and 'science' of cattle breeding, Baldrey (1911: 02) states:

> There is no doubt that Rajputana is an excellent cattle-breeding country. It is sandy and water is scarce, which necessitates cultivation being at a minimum, this leaves huge tracts of country on which, although herbage is sparse and the feed somewhat scanty, the grazing quality is excellent. To the ordinary observer passing through the country, it would appear from its desert-like aspect that rearing cattle would be an impossibility, but if enquiry be made and the cattle in and around the villages be examined, it will be noticed that they are in good condition although their food consists of nothing but what they pick up on the apparently dried up grazing ground.[9]

This statement by Baldrey captures a crucial fact, little understood by many other colonial experts. The best indigenous breeds of draught bullock emerged on tracts that to an unfamiliar eye appeared as 'overstocked and overgrazed village grazing land'! The nomadic cattle breeders who inhabited such tracts reared cattle and then travelled with their year-old cattle to areas such as Kheda, West Punjab etc. to supply the plough and cart bullock needs of cultivators. Baldrey could understand and appreciate mobility when he outlines the contours of nomadic movements stating:

> The people of the Nagaur district as a rule, travel south and west into Malwa and so far as the district around Mhow or into North Guzerat and Sindh. These people are commonly known as *Banjaras*, and it occasionally happens that there is some friction between them and local inhabitants over the grazing, they are, however, generally taxed for their grazing rights.[10]

Another contemporary account by Joslen (1905: 09) further illustrates the mutually reinforcing interrelations that obtained between settled cultivators and nomadic cattle breeders at the turn of the century:

> The Ahmedabad and Kaira districts are well suited for cattle breeding, fodder crops and grazing being usually plentiful. In nearly all villages, a few professional cattle breeders, *Rabaaris* and *Bharwaads* are to be found, and these men may be said to take the place of weekly cattle fairs such as are general in the principal villages of the Deccan. Bullocks for agricultural purposes are reared from the time they are a few months old, at which age they are purchased after castration, from *Rabaaris* and regular buyers and sellers, *Bharwaads*, many of whom go to the Palanpur and Radhanpur States, where the best Kankrej cattle are to be found and buy these calves. Most of the fine cattle in the Charotar[11], Kaira district is thus obtained. At Nadiad, I met a *Rabaari* with 18 Kankrej calves he had brought down from Radhanpur and at Borsad a dealer was present with 15 young Kankrej bullocks and bulls brought by him from Patan in North Gujarat via Viramgam and Dholka, the names of other places visited *en route* he could not give or would not. This is his regular annual tour. Rabaaris who now own from two to three cows up to twenty, used to have from 50 to 150 or 200, with from one or three bulls, before late succession of famine years, with influx of cattle from other parts, nearly resulted in the extermination of breeding stock (cows and bulls), which are almost entirely maintained by grazing. Most of the best agricultural cattle were saved.

Thus, in multiple narratives, we come across a recognition of the indigenous and traditional knowledge of professional and nomadic cattle breeders by experts who worked with Civil Veterinary Department in Bombay Presidency. Not only did they reflect a better understanding of the role traditional cattle breeders played, but these opinions also narrate the relationship that existed between nomadic cattle breeders such as *Bharwad* and *Rabaari* and settled cultivators.[12]

Another animal husbandry expert, K. Hewlett, who was Superintendent of Civil Veterinary Department of Bombay Presidency, notes that "as a rule, the cultivator prefers to purchase bullocks as yearling and rear them himself". Hewlett recounts that professional breeders—*Rabaris* of North Gujarat used to bring large number of young bullocks from breeding grounds to the highly cultivated tracts and cultivators preferred to buy the bullocks as yearlings, rear them well on a nutritious diet and put them to light work when they turn three to four years and they were not put to harder work until five years, when they attain maturity.[13] These narratives also described hardships and sufferings the nomadic cattle breeding class had to undergo, following successive droughts and famines in 1890s.

Furthermore, Baldrey (1911: 05–06) criticized the excessive attention that horse breeding had received in colonial animal husbandry policy and states without mincing his words:

> India being essentially agricultural country and cattle being so important to that industry, is it not a little extra ordinary to find that horses which really do not form a part of agricultural stock of the country and are therefore of secondary importance to the farmer, have received

far more attention than has cattle breeding? Many lakhs of rupees have been spent on the encouragement of the breeding of horses but little on cattle.

Although published in 1911, Baldrey's book was based upon notes he had written in 1905, just five years after the severe famine of 1899–1900. The notes carry the impression of the great famine and Bradley outlines colonial policy interventions when he recommends, "The formation of cattle breeding studs in all centres where good indigenous breeds are present would be a very beneficial measure". He believed that such efforts "would allow of good typical bulls being bred and available to the people in the event of famine or other causes killing off the village stock".

E. J. Bruen, Livestock Expert with Government of Bombay Presidency refers to efforts made by government towards the improvement of cattle right from the year 1899, in his memorandum submitted to Royal Commission on Agriculture. Initially, bulls were placed in each district after inquiries through the agency of District Local Board. In 1899, the year of the great Indian famine popularly known as *'chhappaniyo'*, the Northcote Cattle Farm at Chharodi was established by Government of Bombay Presidency with an express purpose to save Kankreji breed and was managed by a Preservation Society. 589 heads of cattle were purchased at an average cost of Rs. 49 per head and the society depended on public charity for maintenance. Bruen (1927) in Bulletin No 136 of the Department of Agriculture of Bombay Presidency titled 'Cattle Breeding in the Bombay Presidency' gives a detailed account of how the establishment of this farm denoted the positive role played by Hindu sentiments and their appreciation of the cattle breeding undertaken by colonial government. However, in 1907 the farm was handed over to the Government of Bombay as the funds of the Preservation Society couldn't meet the necessary charges.

In 1907, Government of Bombay Presidency started the European method of giving out premium bulls to well-known landlord class persons. Two areas—one in North Gujarat (Ahmedabad district) and the second in Sholapur district in Deccan—were selected for this scheme. A decade later, the government considered the question of cattle breeding to be sufficiently important to start a new section of the Agricultural Department. A special Deputy Director of Agriculture for Animal Breeding was appointed, and he was put in charge of the ongoing work in districts as well as Northcote Cattle Farm, Chharodi, Ahmedabad district. Until 1919, when the farm was handed over to the Civil Veterinary Department, it was managed with a breeding policy that aimed at breeding for draught quality and cows were never milked on the farm. However, in 1921, 49 of the best cows were taken in hand. The annual milk yield for this herd in first year was 480 lbs and it reached 1330 lbs per animal in 1925–'26.[14]

In 1917, the Government of Bombay had proposed to start two cattle breeding farms for milch cattle, one in Karachi to preserve and improve the Karachi or Sindhi breed of cattle and another at Sholapur. In addition to this, the discussion on budget estimate for that financial year suggested that Government was committed to take some definite action on the report of the committee on Milk Supply to Towns in Bombay Presidency, in the shape of a grant to help to improve Poona's milk supply.[15]

However, in 1919, cattle breeding operations were taken over from the Veterinary Department and were handed over to the Agriculture Department in Bombay Presidency.[16]

There was a sudden shift in the cattle breeding policy in the Bombay Presidency around mid-1930s, when it was decided to discontinue the previous system of dividing the presidency areas into 'A' and 'B' zones. Under the previous system, the Livestock Expert was responsible for breeding in 'A' zone areas, where the cattle were of a reasonably pure type and intensive breeding was done; while 'B' zone areas, which were under the control of the Deputy Directors of Agriculture, included all the areas where the cattle were less pure and where bulls produced in the 'A' zone areas could be satisfactorily employed. In replacement of such a division of labour, breeding control throughout the Presidency was assigned to 'Cattle Breeding Centres' under the control of the Livestock Expert.[17]

In the year 1938–'39, Bombay province reported holding 40 cattle shows as against 27 held in the previous year. Not merely the number of shows was rising, but the participation rate had also witnessed enthusiastic response from cultivators and breeders. As against 1126 animals exhibited in the previous year, 40 cattle shows, held during the year 1938–'39 reported 2148 animals on exhibition. Official statistics reported in the annual Agriculture and Animal Husbandry in India report showed that "as compared with 193 premium bulls stationed at different centres as on 1st July 1938, 278 premium bulls were stationed at different centres as on 1st July 1939".[18] On 1st April 1940, Northcote Cattle Breeding Farm at Chharodi, District Ahmedabad—famous for its breeding of Kankrej breed of cattle since 1917—was transferred to the Institute of Agriculture, Anand.

While writing about the efforts being made in Bombay Presidency by Department of Agriculture on the subject of cattle breeding, Keatinge (1921: 119) stated:

> Amongst the millions of *nondescript* cattle to be found in the Bombay Presidency, five excellent breeds stand out prominently and steps are being taken to breed pedigree stock of each distinct breed. Kankreji cattle are being bred for draught purposes and for milk production, Karachi cattle for milk and Mysore cattle for draught. In the near future it is proposed to deal with the Khillari and Krishna valley breeds. Considerable progress has been made in the case of Kankrejis, which are an excellent breed of dual utility. Selected bulls so produced are put out with breeders on the premium system and as the number of premium bulls increases the system is bound to improve the village herds. Delhi buffaloes are being bred at Sukkur in Sindh and the breeding of Gujarati buffaloes will shortly be taken up.

The committee on Milk Supply to Towns in Bombay Presidency was formed by the Bombay Government to attempt the solution of the problem of how to increase the milk supply to meet the demand and at the same time keep the prices in check.[19] Within a few years after that, responding to the repeated airing of the complaints and concerns by the public in Bombay Presidency about the rapid deterioration of both the draught and milch cattle, the Government of Bombay appointed a committee-wide Order No. 3252, dated 24th December 1921 to discuss about the cattle breeding question. The committee had following seven members: Rao Bahadur G. K. Chitale (Chairman), Mr. E. J. Bruen (Dy. Director of Agriculture for Animal Breeding Bombay Presidency) (Secretary), Mr. Nariman R. Kothawala, Mr. S. H. Kurli, Mr.

E. M. Hodgson (Conservator of Forests), Khan Saheb J. D. Buxy, (Dy. Superintendent, Civil Veterinary Department, Bombay Presidency) and Rao Saheb Bhimbhai M. Desai, (Dy. Director of Agriculture, Gujarat). The committee met at Poona on 13th March 1922 and spent the next two days examining local experts and witnesses. The committee had articulated a recommendation to be considered a permanent committee on cattle breeding in Bombay Presidency that would meet once or twice a year to discuss and lay out the plans to be followed next year.

The report by committee appointed to review cattle breeding and dairying in Bombay (1923: 02) took note of extinction of the professional breeders who depended entirely on grazing for their livelihood in Kapadvanj taluka of Kheda district and Viramgam taluka of Ahmedabad district, where the *Rabaris* were described to have completely abandoned this profession. The committee also noted that in Deccan, there was a class of persons known as *Gowsavi*, who at one time took away dry cows and buffaloes and grazed them on the vast areas of grazing land then available, but who were finding it increasingly impossible to continue this practice, owing to grazing restrictions and the very rapid growth of cultivation.

The Committee had visited four cattle breeding farms—namely Chharodi, Bankapur, Surat and Tegur—in the Presidency. The Committee (1923: 04) after hearing the evidence of several persons, both cultivators and breeders, therefore recommended to Government that certain areas well suited for cattle-raising such as Kapadvanj in the Kheda district and Jambusar in the Broach district and similar areas in other breeding tracts, be specially assigned to the raising of cattle by professional breeders. This recommendation of the Committee had come out of its intent to prevent discontent between the cultivator and the breeder and with a view to ensure supervision by the Cattle Breeding Department. In this report there are already signs of how the diminishing grazing lands and expansion of cultivation, under the extensification strategy had played a crucial role in accentuating the social stratification further.

The Committee (1923: 05) categorically stated that it does not recommend introduction of any foreign blood from outside India and recommended that local, indigenous pure breed be encouraged. This view was reiterated by Bruen at the Cattle Conference next year at Bangalore (1924: 09) when he stated that he was against crossing of any kind whatsoever and expressed his opinion that the present deplorable condition of Indian cattle was due to indiscriminate crossing. Speaking about the cattle breeding work being done by the Government in Bombay Presidency, Mann stated at the same conference (1924: 09) that government had devoted their energies to the development of four different types of Bombay cattle—Kankreji at Chharodi and Surat, the Amrit Mahal at Bankapur, and the Sindhi and Tharparkar at Karachi. Their efforts, he indicated, in developing and distributing these breeds had achieved a measure of success and the demand for bulls and the general interest taken in the cattle question in Bombay had greatly increased.

The Bombay Government issued a resolution in August 1924 on the Cattle Committee's report, thereby "giving general approval to the schemes suggested by the committee for starting new farms, and for the grant of additional financial help to the existing farms, but added that schemes could only be taken up gradually". Four years later the Royal Commission on Agriculture had taken note that the resolution also called for elaborate proposals, with reference to the suggestion that intensive breeding operations should be conducted in select talukas, but came across no definite action on that front.

The annual administrative report for the year 1921–'22 as well as for years 1922–'23, 1923–'24 and 1924–25 shows that while some thinking around the question of interventions in cattle breeding had surfaced, the report showed "Nil" against the subject, "Breeding Operations: Bovine"! In contrast to this, the Civil Veterinary Department's annual administrative reports continued to show the progress with regard to horse and mule breeding work. Annual administrative reports of the Civil Veterinary Department for the years 1925–'26 and 1926–'27 curiously stopped this disarming frankness in showing "inaction" on this front and started to report only about "Horse and Pony Stallions" under the subject, "Breeding Operations". However, this must not be interpreted as if the Bombay Presidency government was not intervening in the cattle breeding at all. This apparent 'inaction' by the Civil Veterinary Department on this front followed from the policy whereby Dr. E. J. Bruen continued to oversee "animal breeding" from his position of being Dy. Director of Department of Agriculture in Bombay Presidency.

Talking about whether cattle breeding and cattle keeping was a subsidiary occupation in addition to crop cultivation, Rao Sahib Bhimbhai M. Desai, Deputy Director of Agriculture in Gujarat and Surat districts told Royal Commission on Agriculture that cattle breeding in 1928 was less than the extent in which it was resorted to ten or fifteen years earlier. When asked to explain the reason for this, he stated that "the areas lying under grass have all been brought under cultivation on account of the high price of cotton and foodstuffs, so that they are not able to maintain the number of cattle they had before".

The Royal Commission on Agriculture (1928: 115) reports figures from census of livestock in Bombay from 1909–'10 till 1924–'25. The figures that it quotes for the number of cattle as reported in previous livestock censuses are: "1909–'10, 7.5 million; 1915–'16, 9.0 million; 1919–'20, 8 million; 1924–'25, 8.5 million". As per the 1924–'25 census, there were over 3.3 million plough cattle, 2.8 million milch cattle and 2.8 million cattle for other purposes. The sharp fall in the number of cattle in 1919–'20 was the result of serious drought in 1918–'19 and 1919–'20. The rapidity of recovery in favourable season is shown by the figures of 1915–'16 and 1924–'25. Royal Commission on Agriculture (1928) noted that "there are at present, 11 plough cattle and 9 milch cattle for every 100 acres cropped". Speaking about ovine population it said, "the number of sheep and goats in 1924–'25 was 1.8 million and 2.8 million respectively".

Talking about the work being done by the Department of Agriculture on 'cattle breeding and dairying,' the Royal Commission on Agriculture (1928: 127) states, "Bombay (Presidency) has many breeds of cattle, each suited to its own region, of which the following eight have distinctive characteristics: Kankrej in Gujarat, Malvi in Khandesh, Krishna Valley in Belgaum, Amrit Mahal in Dharwar, Gir in Kathiawar, Dongri in Thana and Nasik, Khillari in the Deccan and in Sind Tharparkar for draught and Karachi for milk." The Royal Commission on Agriculture (1928:

217) notes about three cattle breeding farms in Bombay Presidency: Chharodi, in North Gujarat district where about 200 cows of the Kankrej breed are maintained on some 2300 acres; Bankapur, in Southern Mahratta country, which maintains a herd of 50 Amrit Mahal breed and Philai, near Karanchi, which maintains a herd of Sindhi cattle, then known as one of the best milking breeds in India. It reports that, in addition to these farms, Presidency government was also subsidizing two *gowshalas* to produce pedigree Gir cattle. Some other *Gowshalas* have also been recognized and were being conducted under the Government advice. Some half a dozen cattle shows were organized each year. The Royal Commission on Agriculture (1928: 217) states that "in Bombay Presidency, the subject of milk supply had received much attention, from those responsible for cattle improvement, partly due to the fact that Bombay city creates a large demand for milk" and was then being provided with a very poor supply of it and "partly because North Gujarat produces large quantities of buffalo milk and at the same time, maintains a good class of draught cattle". While the Royal Commission on Agriculture (1928: 217) voices a poor opinion on the grazing land generally, it was in North Gujarat that they witnessed management of grazing land that was akin to what was practised in Britain, where farmers practised enclosures and "the wide grass borders of the enclosed fields provided the only approximation to "pasture" in association with tillage land to be seen in India" which in turn rendered "the lot of the cultivators' favoured animals, she-buffaloes and draught oxen relatively a happy ones".

Talking about the breeding policy practised in Bombay Presidency, the Royal Commission on Agriculture notes that it revolved around three methods: "(i) to breed more and more milk into each breed, (ii) to breed for early maturing qualities and (iii) to breed regularity of calving into stock of bulls it put out.[20] It also took a note that "the double demand in Gujarat for good plough and good milking cattle has, naturally enough, suggested the desirability of combining draught and milking qualities in one breed of animal". This must have led to witnesses who provided Commission with evidence and appeals, "urge upon the need of replacing buffaloes by good cows of a 'dual purpose' type". Such requests from North Gujarat cultivators for 'dual purpose' cattle and to "replace buffaloes"—although the livestock holding pattern in Kheda suggests a trend for preferring buffalo as principal milch animal—were possibly rooted in discursive strategies of identity formation vis a vis a colonial State. One needs to remember that the native elites who gave written and oral evidence before the Royal Commission on Agriculture representing the cultivators were responding positively to the then prevalent cow protection discourse, whether emerging from religious formations or coming from other humanitarian perspectives. The anti-colonial struggle of that time had woven the cow protection discourse within the articulation of its understanding of village-centric economy and programme for rural development by positing at discursive level Cows versus Buffalo polarization.

The Royal Commission on Agriculture (1928: 218) reports that the Bombay Cattle Committee (1923) had studied in detail, the issue of distribution of pedigree bulls in Bombay Presidency. They reached the conclusion that "the improved bulls from government cattle farms should be located in a series of talukas" which were selected by the Committee as being natural breeding areas. The Committee made a plan "to locate a cattle breeding farm in each track, stock it with the best cows available, and provide it with selected bulls from the central cattle breeding farms and breed young bulls for local issue". The farm manager was supposed to "exercise general supervision over the bulls placed out within the selected areas, follow the progress of the progeny and compile a register of all good cows in his district".

It is important to underline how E. J. Bruen, Livestock Expert to Government of Bombay talked about the situation of professional cattle breeders in the year 1927:

> The professional cattle-breeder is fast dying out, the only professionals now remaining are the Rabari and Bharwad of North Gujarat and the Dhangars of Ahmednagar district. But the cattle-raising is now generally confined to villages, the cattle are kept more or less constantly in one place, the treatment given is nevertheless practically the same as under nomad conditions, i.e. the cattle are not fed and dry cows are expected to thrive on the grass available (which is very meagre for six months of the year). In the grazing fields, moreover, as many adult entire males are found as females and cattle are continually mixed. When disease appears, therefore, it takes a toll of 30 to 40 percentage, and those that do recover are usually unfit for further use as breeding animals. The net result is under-feeding, promiscuous in-and-in-breeding, crossing and the constant dread of diseases. Naturally under such conditions, a deterioration has set in, which will be difficult to check.

These words by Bruen occur in his 'Note on Cattle Breeding' submitted to Royal Commission on Agriculture after a paragraph that has a nostalgic tinge of how "a professional cattle breeder raised his cattle by migrating from place to place in search of food and water for his cattle". The pastoral mobility that was *sin qua non* for a professional cattle breeder was restricted when colonial authorities passed forest governance rules that prohibited their movement. While it is not clear how effective and how widespread were the prohibitory orders passed by Veterinary Authorities during the time of cattle plagues, upon reading annual administrative reports of Civil Veterinary Departments, one does get a feeling that they often considered restricting wandering herds and segregation as the first action for stamping out a contagious disease.

There was yet another shift in the cattle breeding policy in the Bombay Presidency around mid-1930s, when it was decided to discontinue the previous system of dividing the presidency areas into 'A' and 'B' zones. Under the previous system, the Livestock Expert was responsible for breeding in 'A' zone areas, where the cattle were of a reasonably pure type and intensive breeding was done; while 'B' zone areas, which were under the control of the Deputy Directors of Agriculture, included all the areas where the cattle were less pure and where bulls produced in the 'A' zone areas could be satisfactorily employed. In replacement of such a division of labour, breeding control throughout the presidency was assigned to 'Cattle Breeding Centres' under the control of the Livestock Expert[21].

4.4 Livestock Improvement Act, 1933

Another development that occurred in Bombay Presidency in 1933 had consequences for the cattle breeding. This was the legislation Livestock Improvement Act (1933), which provided for the compulsory castration of inferior male stock in villages on the initiative and at the instance of local bodies and extended the authority and power of the veterinary department. Provided that local bodies agreed for the application of this legislation, the veterinary officers gained power to order compulsory castration of all bulls in the area, except those that have been licenced for breeding.

Rules under the Act appear to have been published in the year 1935 and their application was taking place gradually.[22] Bombay Presidency was a pioneer in legislating such an intervention and the second meeting of Animal Husbandry wing of the Board of Agriculture and Animal Husbandry (1937) discussed the provision of the Act and debated the practicability to introduce those provisions in other parts of India. In 1935, there were only two areas that were brought under the ambit of the Bombay Livestock Improvement Act, but by 1942, the act had become applicable in 73 villages and some more villages were under the survey. By the year 1941–'42, the application of this act covered 77 villages in the province and "2903 castration of scrub bulls were carried out by Veterinary Assistant Surgeons". By 1939–'40, a similar bill on the lines on Bombay Livestock Improvement Act was to be introduced in Madras and the similar work was being done in several villages of Bhavnagar State.

Summarizing developments on Animal Husbandry, subsequent to the submission of the report by subcommittee on Animal Husbandry and Dairying in the year 1939–40, which was part of the National Planning Committee endeavour, Prof. K. T. Shah (1948: 109) observed that the Bombay Livestock Improvement Act, 1933 that intended to prevent of undesirable bulls or male stock, was by then adopted by 112 villages and it needed to be extended on much larger area.

During the year 1940–'41, under the provisions of this act, 7600 castration of scrub bulls were done with the help of Veterinary Assistant Surgeons. During the year 1941–'42, seventeen more villages had been surveyed with an objective of applying this act.[23] By the year 1944–'45, the application of this act covered 92 villages in Bombay province.[24] By the year 1950–'51, the application of this act covered 218 villages in Bombay state and proposals for the application of the act to 83 more villages were under the consideration of the Government.[25]

4.5 Viceroy's Appeal to Zamindars for Gift Bull Scheme

In 1937, the Viceroy Victor Hope appealed for donation of bulls for the campaign on cattle improvement.[26] Proceedings of the Second Cattle Conference (1938: 81) reports that "some difficulties arose in Bombay, since some donors desired to earmark their gifts for particular villages in a manner incompatible with any definite scheme for livestock improvement" and attempts were being made to resolve this as well as "to create a Central fund to be placed at the disposal of the Livestock Expert for use for cattle improvement".

4.6 Cattle Breeding in Gaekwad-Ruled Baroda State

With respect to the condition of cattle in Gaekwad-ruled Baroda state in the mid-1920s, Amritlal C. Desai, the then Director of Agriculture stated the following opinion:

> The question of milk and cattle breed, so far as Guzerat and Kathiawar are concerned, is intimately connected with the question of fodder. Every third or fourth year, there occurs a drought resulting in the loss or destruction of a large number of cattle. The farm economy in such years is so much upset that it takes two or three years before the normal condition is reached. The primary question for dairy or cattle breeding is the preservation of fodder. The number of cattle is fairly sufficient. The milk supply in all the big and small towns deficient and pure and unadulterated milk is rarely to be obtained.[27]

Well-to-do cultivators in Baroda state of the mid-1920s used to keep two cows and two she-buffaloes in addition to a pair of bullocks for ploughing and cart pulling. Except poor *Kolis* and *Dublas*, most cultivators owned their oxen and buffaloes. Large herds of cows and buffaloes were maintained by professional herdsmen or *Rabaris*, who used to sell their produce in the shape of clarified butter or *ghee*. The number of milch cattle in Kadi division of the Baroda state was larger than any other part of Gujarat and the breed was held to be superior. Cultivators in the far south division of Baroda state such as Navsari used to buy large powerful oxen (known as *Hedia*) from travelling herdsmen from Kadi and other parts of North Gujarat.

In his written submission and oral evidence before the Royal Commission on Agriculture, Govindbhai H. Desai, the Naib Deewan of Baroda state mentioned that there were several cattle breeding stations for upgrading the indigenous breeds but many were closed down due to lack of interest on the part of cultivators.[28] In 1927, there was only one cattle breeding station at Makarpura, near Baroda and bulls and buffaloes for the use of the cultivators were supplied to the local boards, on their undertaking to maintain them and to regulate service.

A review report by the Imperial Council on Agriculture Research on Animal Husbandry for the period 1934–'35 and 1935–'36 (1936: 16) states that 110 animals were maintained at Makarpura cattle breeding farm at the end of 1934–'35 and to help the general improvement of agricultural cattle, a number of selected bulls were distributed on the premium system. A similar review report for the year 1938–39 by ICAR (1941: 244–245) stated that some nine Kankrej bulls were at stud, as against the three during the previous year. The report also took notice of six Gir bulls from high milking strains placed out as stud bulls.

4.7 Seeking Cooperation from Gaushalas for Cattle Breeding

Speaking at the fourth meeting of Animal Husbandry wing of Board of Agriculture and Animal Husbandry, Mankar recalled that at the instance of Leslie Wilson, Governor of Bombay, a Cattle Improvement and Dairy Farming Association was started in the Bombay Presidency in the year 1929 by the efforts of livestock expert, E. J. Bruen. Attempt was made to start district associations and one was started at Poona. Village associations were also started and propaganda was carried out with the help of a touring car and the objects of association were explained to the cultivators. Village committees were formed for helping in the maintenance of the stud bulls for improvement of cattle.[29] Later, the cooperation of *Gowshalas* was also secured. The Imperial Dairy Expert, W. Smith suggested that such cattle breeding cooperative societies "shall inaugurate milk recording scheme, whereby milk yields of the best cows and buffaloes belonging to the members of the society could be recorded and authenticated". Smith was hopeful that "if a cooperative society honestly worked, the value of the cattle and the yield of milk could be doubled in three generations, i.e. ten years".

Since the year 1924–'25, a cow protection group, the Bombay Gowrakshak Mandali (established on 23rd January 1889) was seen cooperating with the Livestock section within the Department of Agriculture, with an aim at improvement of the Gir cattle at their two farms at Kandivali and Betegaon. This cooperation between Gowrakshak Mandali continued for around two decades for "the improvement of the valuable Gir breed of dairy cattle". On 23rd January 1939, the Mandali celebrated its Golden Jubilee by organizing a large Cattle Show (where 371 Gir cattle were exhibited) and inviting the Viceroy, Lord Linlithgow.

Another Gowshala in Bombay city, namely Nathu Lalji Charity Trust Fund Gowshala had also came forward to cooperate with the Livestock Expert to Government of Bombay and was working on similar lines to breed Gir cattle since the year 1939.[30] Similar instances of enlisting the support of cow protection groups were witnessed in Odisha, where 13 cattle shows were held in the year 1938–'39, under the auspices of the Utkal Gomangal Samiti and two animal welfare shows in Cuttak and Berhampur towns were held by the SPCA".[31]

By 1941–'42, there was a feeling amongst officers of Livestock section within the Department of Agriculture, Bombay province that all the numerous Gowshalas and Pinjrapoles operating in the districts should cooperate with the department. In the previous year, the Livestock Expert had paid personal visits to Gowshalas at Barsi, Dhulia, Nandurbar and Sholapur and had discussed with their management the advantage of cooperation with regard to livestock improvement work, but there was no reciprocation.[32] The annual report of the department 1941–'42, underlines this feeling by voicing the frustration that "[I]nspite of continued propaganda work by the Livestock Expert and frequent discussions with the management of numerous Gowshalas and Pinjrapoles in the districts, *there is not much progress to be reported in this direction*". The newly appointed Livestock Expert to Government of Bombay Province stated in his report on the works carried out during the year 1941–'42[33]:

> Almost all the Gowshalas and Pinjrapoles in the province were revisited for inducing them to cooperate in the improvement of cattle, *but so far no institute is willing to cooperate.* Further attempts would be made. (emphasis added)

This alleged lack of enthusiasm on the part of Gowshalas Pinjrapoles management to cooperate with the Livestock Expert on cattle breeding and livestock improvement work must sound curious, given that for around two decades by 1941–'42, two major Gowshalas in Bombay city had continued to work as allies of the Department of Agriculture on improving the milking strain in Gir cattle. The legislative route was actively being proposed by early 1940s "to control the working of these institutions in the interest of livestock development".[34] During the year 1944–'45, the Livestock section within the Department of Agriculture, Bombay province succeeded in enlisting the support of three Gowshalas at Bijapur, Sholapur and Bagalkot.[35]

By the year 1938–'39, there existed around five hundred gaushalas or pinjrapoles in India. In official statistics and reports on the running of these institutions, only sixteen of these institutes were accounted for and they were reported to have housed over seventeen thousand animals during the year. The total expenses incurred by them were said to be Rs 5.5 lakh, which amounted to Rs 33 per animal.[36]

4.8 More Focused Research on the Economy of Cattle Breeding Tracts by ICAR (1937 Onwards)

In 1937, the Imperial Council of Agricultural Research also undertook an inquiry in seven prominent breeding tracts to determine milk yields. From 1939, this inquiry was followed by a systematic programme of defining the characteristics of various breeds. Following this, ICAR published a catalogue of 25 breeds of cattle and 7 breeds of buffaloes that showed promise. Further a pedigree registration was taken up in 1941 with respect to the breeds of all-India importance and a national cattle breeding policy was announced. The main feature of this national cattle breeding policy, once more underlined the aim to develop dual purpose cattle by increasing the milk-yielding capacity of the indigenous breeds without impairing their draft quality. Following this, a long-range research programme to evolve the dual purpose *Hariana* cattle was started in 1941 in Punjab and in the following year, a similar programme with respect to *Kangayam* breed was undertaken in Tamil Nadu.[37]

Notes

1. For more on the animals used for military campaigns see, Sivasundaram, Sujit (2005) 'Trading Knowledge: The East India Company's Elephants in India and Britain', *The Historical Journal*, Vol 48, no 1, pp. 27–63.

2. A study on *The Organisation of Government of Mysore*, mentions that "The State had at one time 240 Amritmahal Kavals covering an area of 4 lakh acres and maintaining 60,000 heads of Amritmahal breed cattle. With the diminished use of this marshy breed of cattle for military purposes and increased use of machines for cultivation purposes, Government reduced the extent of Amritmahal Kavals and the number of cattle. At the end of 1967, there were only 64 Kavals with a total extent of about 70,000 acres and roughly about 2000 heads of cattle". See, Rebello, R. J. (1973) *The Organisation of Government of Mysore*, Indian Institute of Public Administration, Mysore Regional Branch, Bangalore p. 70. This breed's history is fascinating and requires more scholarly attention. W. D. Gunn in his book, *Cattle of Southern India* (1909: 12–19) devotes several pages to discuss the military importance of this breed and cites Captain Davidson's report on the Amrit Mahal cattle attached to the Bombay column of the English Army in Afghanistan in 1842, thus: "No draught cattle in either army were so efficient as the 230 Mysore bullocks which accompanied the Bombay troops to Afghanistan. It was entirely due to this very superior description of cattle that no part of Bombay Park was required to be abandoned when the troops were returning to India over the almost impractical roads through the Tirah Mountains. These cattle were frequently upwards of sixteen hours in yoke. The draught bullocks of the Bengal Army were the property of the Government, and were not in my opinion, as fine animals as the Mysore bullocks". Providing some signposts in the history of this breed, Gunn (1909: 12) states that "In 1813, the Amrit Mahal cattle along with the pasture lands were handed over to Captain Harvey of the Madras Commissariat...In 1840, the Maharaja's Amrit Mahal herds and grazings were amalgamated with those of the British Government, and the whole placed under the orders of the Mysore Commission...In 1860, from the motives of economy, Sir Charles Trevelyan ordered the establishment to be broken up, and the herds to be sold...In 1866, efforts were made to re-establish by the cordial approval and assistance of the late Maharaja... However, while the late Maharaja was a large purchaser when the old establishment was broken up by six years earlier, very few were obtained owing to the Pasha of Egypt having secured most of the best blood... and thus in 1870, the Madras Government was able to obtain sufficient stock (4000 cows and 100 bulls) to fairly start again...In 1883, the British Government handed over this valuable property to the Government of His Highness Maharaja of Mysore for two and a quarter lacs and agreement to supply to Madras Government 200 bullocks annually". Gunn had mentioned existence of 210 kavals and translated it as "reserved pastures". From Voelcker's *Report on the Improvement of Indian Agriculture* (1893: 198), we come to understand that while the heard was broken up in 1885 (sic) a certain number of cattle were kept by the Mysore Government as Hosur and the Bhadgaon (Bombay Presidency farm) herd owed its origin to this.
3. See, Morgan, J. W. A. (1907) 'The Government Cattle Farm, Hissar, Punjab', *The Agricultural Journal of India*, Vol II, Part iv, pp. 364–373. For a longer narrative on how the interest emerged in camel and cattle breeding around the period 1809–1815 see, Caton, Brian P. (2014) 'The Imperial Ambition of Science and its Discontents: Animal Breeding in Nineteenth Century Punjab', in Rangarajan, Mahesh and K. Sivaramakrishnan (eds) *Shifting Ground: People, Animals and Mobility in India's Environmental History*, Oxford University Press, Delhi, pp. 132–154. Also see, Caton, Brian P. (2013) *Writing the History of a Colonial Institution: The Case of the Government Cattle Farm, Hissar*; NMML Occasional Paper No 38, History and Society (New Series), Nehru Memorial Museum and Library, Teen Murti House, New Delhi. Caton refers to probably the only account of the Government Cattle Farm, Hissar in print written by W. S. Read in 1937. See, Read, W. S. (1937) *The Government Cattle Farm, Hissar: A History and Guide*, Superintende, Government Printing, Punjab, Lahore. In 1890, Hissar Farm was spread in an area that measured 67 square miles and housed 7000 heads of cattle as reported by Voelcker (1893: 201). Voelcker had reported that from 70 to 80 bulls were sold annually for agricultural purposes at the Government price of Rs 150 each, while 350 heads were supplied yearly to the Commissariat Department. Not only did Voelcker cite from the annual reports from surrounding districts to impress on his readers how much good stamp was being left by the Hissar bulls supplied from the Government Farm, he also underlined

that "in 1887, two Hissar bulls were sent as far as Arrah (Behar) for use on the Government Estates there".
4. During his tour, Voelcker had visited two cattle farms, Hissar in Punjab and Bhadgaon in Bombay Presidency. He however mentions other Government Cattle Farms such as the Saidapet Farm in Madras and the Cawnpore Farm in North West Provinces.
5. Department of Agriculture, Punjab (1910) *Cattle and Dairying in Punjab*, Department of Agriculture, Punjab, p. 05.
6. Mollison J. and L. French (1907) 'The Montegomary and Sind Breeds of Cattle', *Agricultural Journal of India*, Vol II, No 2, p. 255.
7. The Annual Report of the Civil Veterinary Department of Bombay Presidency for the year 1899–1900, p. 13.
8. Quoted in The Annual Administrative Report of the Civil Veterinary Department of India for the year 1899–1900, p. 84.
9. Bradley, F. S. H. (1911) *The Indigenous Breed of Cattle in Rajputana*, in Breeds of Indian Cattle Series, Civil Veterinary Department, Government of India Press, Shimla. It is important to note that Capt. F. S. H. Baldrey had made notes on the breeds of cattle found in Rajputana during his travels in the country in the Autumn and Winter of 1901 and 1902. While reading his remarks, thus we need to keep in mind that he was looking at Rajputana immediately after the Great Famine memorized in popular narrations as Chhapaniyo (the 1899–1900 famine). Baldrey (1911: 05) notes with a lamenting tone that in Balotra and Bikaner districts, known to be headquarters for breeding of Nagori cattle, only 5–10% of the cattle survived (the great famine) and the owners of these even were travelling in Sindh and other places seeking fodder.
10. Bradley, F. S. H. (1911) *The Indigenous Breed of Cattle in Rajputana*, in Breeds of Indian Cattle Series, Civil Veterinary Department, Government of India Press, Shimla.
11. Charotar region of the erstwhile Kheda district consisted of four talukas, namely Anand, Borsad, Nadiyad and Petlad. Note that Petlad in the late colonial period was a sub-division governed by Gaikwad's Baroda State.
12. For an account of Rabaari and Bharwad cattle breeders from Kankrej, see Patel, M. D. and C. N. Dave (1944) 'Kankrej Husbandsmen of Gujarat', *Indian Farming*, Vol V, No 5, (July) Pp. 276–278.
13. Hewlett, K. (1879) Breeds of Indian Cattle: Bombay Presidency, Government Printing Press, Calcutta.
14. This paragraph on the cattle breeding at Northcote Cattle Farm, Chharodi draws upon information contained in the Memorandum on Cattle Breeding in Bombay Presidency submitted by Mr. E J Bruen to Royal Commission on Agriculture. See Royal Commission on Agriculture, Vol II, Part I (Evidence taken in Bombay Presidency), pp. 399–437.
15. Editorial in The journal of Dairying and Dairy Farming in India, Vol IV, Part 3, p. 145.
16. The question about which department is better endowed to control cattle breeding operations had remained a controversial topic for discussions at the meeting of the Board of Agriculture and at the meeting of Veterinary Officers in India as well as at Cattle Conferences. Such abrupt changes in the control also meant that sometimes the advances gained by a specific cattle breeding policy got lost with the change of guards.
17. Agriculture and Animal Husbandry in India for the years 1933–'34 and 1934–'35, p. 09. This report also tells us that while "the purpose of the Department (as per cattle breeding policy adopted for Bombay Presidency) was to breed a dual purpose breed of cattle", it was witnessed that "a steady progress was made in improving the milking capacity of the Kankrej herds at Chharodi and Surat". It was also reported that "all the cattle farms (under the Bombay Presidency) showed that by systematic selection, breeding and proper care, earlier maturity and greater regularity in calving could be ensured".
18. Agriculture and Animal Husbandry in India for the year 1938–'39, p. 275.
19. Main recommendations of the committee report were discussed by Meolcian in an article published in The Journal of Dairying and Dairy Farming in India, Vol III, Part 1, p. 37.
20. A report by Imperial Council of Agriculture Research that reviewed the period 1934–'35 and 1935–'36, stated that the Bombay Presidency continued with the policy to breed 'dual purpose'

animal, while simultaneously attaining steady progress in improving the milking capacities of the Kankrej herds at Chharodi and Surat. The review notes that "reports of all the cattle farms showed that by systematic selection, breeding and proper care earlier maturity and greater regularity in calving can be ensured".

21. Agriculture and Animal Husbandry in India for the years 1933–'34 and 1934–'35, p. 09. This report also tells us that while "the purpose of the Department (as per cattle breeding policy adopted for Bombay Presidency) was to breed a dual purpose breed of cattle", it was witnessed that "a steady progress was made in improving the milking capacity of the Kankrej herds at Chharodi and Surat". It was also reported that "all the cattle farms (under the Bombay Presidency) showed that by systematic selection, breeding and proper care, earlier maturity and greater regularity in calving could be ensured".
22. Agriculture and Animal husbandry in India 1934–'35 and 1935–'36, ICAR, 1936, p. 09. For the copy of the Act and Rules see, Proceeding of the Second Meeting of Animal husbandry Wing of Board of Agriculture and Animal Husbandry, pp. 112–119.
23. Annual Report of the Department of Agriculture, Bombay Province, 1941–'42, p. 105.
24. Annual Report of the Department of Agriculture, Bombay Province, 1944–'45, p. 23.
25. Annual Administrative Report of the Department of Animal husbandry and Veterinary Science, Bombay, 1950–'51.
26. Victor Alexander John Hope, the second Marquess of Linlithgow, was chairman of the Royal Commission on Agriculture (1927–'28) and held the office as Viceroy of India from April 18, 1936, to October 01, 1943.
27. Cited in Chatterjee, Nilkanth (1926) The Condition of Cattle in India, All India Cow Conference Association, Calcutta. p. 361.
28. The proceedings of the fourth meeting of Board of Agriculture in India (1909: 36) had approved the programme submitted to it by M. A. Sitole, then Director of Agriculture of Baroda state, which mentioned that efforts were being made to introduce good milking strains of cattle and to improve local breed by crossing it with half English Gujarat bull. The Baroda Model Farm at that point of time was already in the process of distributing young bulls of its own breeding to cattle breeders in the district. The programme submitted to the third meeting of Board of Agriculture (1909: 63) had reported that the Baroda State was not only trying its hands at crossbreeding, it also had constructed a Cheese House as early as in 1908, for making cheese and utilizing Skim Milk.
29. The Imperial Dairy Expert, William Smith had penned a note on 'Cattle Improvement Cooperative Scheme' on the request of Gandhi, when Gandhi visited Imperial Institute of Dairying and Animal Husbandry, Bangalore in the year. Extract from the same was published by Gandhi in Young India dated 04-08-1927.
30. Annual Report of the Department of Agriculture, Bombay Province, 1941–'42, pp. 33–34.
31. Agriculture and Animal Husbandry in India for the year 1938–'39, p. 277. The report also underlines that the Department of Agriculture in Odisha continued to extend a grant of Rs 1500 to Utkal Gomangal Samiti as usual.
32. Annual report of the Department of Agriculture, Bombay Province, 1940–'41, pp. 37–38.
33. It is important to note that Edward J. Bruen, O.B.E, I.A.S. who served as livestock expert to Government of Bombay for more than two decades had died suddenly on 7th December 1941 and D. B. Gokhale was appointed as Livestock Expert on 9th December 1941.
34. Annual Report of the Department of Agriculture, Bombay Province, 1941–'42, p. 37.
35. Annual Report of the Department of Agriculture, Bombay Province, 1944–'45, p. 23.
36. Agriculture and Animal Husbandry in India for the year 1938–'39, p. 183.
37. See, Randhawa, M. S. (1986) A History of Agriculture in India, Vol IV (1947–1981), ICAR, Pusa, Delhi, p. 545.

Chapter 5
Writing the Pre-history of Amul: Emergence of Urban Milk Markets in Late Colonial India

This chapter examines historical narratives and processes that shaped the evolution of milch cattle and organized dairying. It explores how the privileging of the food production function of cattle was closely linked with the question of city milk supply. This chapter tries to understand how rapid urbanization and rising milk prices in urban areas had led to emergence of a specialized form of dairy farming in the form of city milch stables. However, the problem that was posed by such a specialized form of dairy farming in the form of city milch stables was the way it enabled higher rates of slaughter of youngstock and hence it came under criticism from the cow protectionists in late nineteenth-century India. This chapter also engages with the way adulteration of city milk supply and unsanitary methods of handling of milk and milk products in bazars prevailed in Indian towns and cities. In those early narratives, adulterated milk was said to cause diseases and hence there was a stress upon the need to have stringent regulations to be enforced to ensure the supply of pure milk.

One such concern finds an articulation in the year 1894, when P. W. O'Gorman read out a paper at the Indian Medical Congress which strongly called out for enactment of legislation to ensure the purity of the city milk supply:

> If there be any article of diet in India that needs more than any other, the earliest and most stringent regulation to be enforced against disease and adulteration, I claim them for dairy produce…Time after time is disease traced to this produce; lives are being sacrificed every day, and outcries arise from every station in India against the present regime. The time is, therefore, ripe for legislation. All European countries have laws in force against adulteration of food, and the last were passed by the Senate of the United States in 1892, under the name of the 'Pure Food Bill'.

The paper also voiced a scathing criticism about the unsanitary conditions of cow-byres in Calcutta city milch stables and reminded his audience that in the year 1886, Dr. Kenneth McLeod of Calcutta, in his lecture before the Calcutta Health Society had underlined the need to address the problem of "revolting condition of the squalid *goala barees* of the city and suburbs". The Calcutta Health Society had in turn sought legislative help from the Government of Bengal, but to no avail.[1] Similarly,

H. Upadhyaya, *Critical Insights on Colonial Modes of Seeing Cattle in India (1850–1980)*, Asia in Transition 27, https://doi.org/10.1007/978-981-97-1560-2_5

O'Gorman paper described in vivid and picturesque details the cruelty exercised by *goalas* engaged in city milk supply and talked about how members of Society for Prevention of Cruelty against Animals had sought legislation to make cruel practices in milking such as *phooka* a punishable offence.[2]

The pronounced contempt for native ways of handling milk and an advocacy for improving their ways by introducing the English or Continental style of city milk supply remained a recurrent theme right from the 1890s till the first two decades of the last century. Thus, a book, addressed mainly to a *gora sahib* of British India, titled *The Indian Amateur Dairy Farm* by Landolicus (1895: 01–02) had voiced the following condescending words in its opening paragraph:

> The few 'Dairy farms' properly speaking that there are in India, are chiefly confined to towns, and these are principally in the hand of natives, managed in a very unsatisfactory way, and their products much adulterated, so much so in such a way as not only to be repugnant, but they are also deleterious to health, and a frequent cause of the spread of diseases of most serious character, such as typhoid fever, cholera, small pox etc... It is true that English modes of treating cattle are not entirely suitable to this country, but at the same time there is much in the practice of treatment of cattle in India which should be avoided, and the English and Continental system followed as nearly as possible. The introduction of a new system to the *goalah* or cowherd caste in India seems one that he abominates and abjures; but let me ask, what new ways are ever pleasing to the natives of this country? Any innovation is always fought against and shirked, whether in gardening, the keeping of a poultry yard or cows... It is well known that *goalahs* will do as natives always will, and that is to follow, the same lines as their forefathers from time immemorial.

These sentiments that the milk being supplied by vendors or procured by *khansama* for a *Gora Sahib's* kitchen was so unreliable and prone to adulteration had reached such an alarming proportion that a few European women thought it fit to share their experience of keeping cows with the community.[3] Sometimes such an author was not even a trained veterinary medicine practitioner (unlike Isa Tweed who combined her knowledge of veterinary medicine with homoeopathy in her book), but took upon herself to convince her people that "a good many tricks of the native are the chief obstacles to cow keeping in India (by their European masters)" and hoped that after reading her book if any young housekeeper learns "how to keep cows so as to combine comfort to her family with profit to her purse", her efforts wouldn't go in vain.[4]

A report on 'Cattle and Dairying in Punjab' published in the year 1910 described the unsatisfactory condition of the milk supplied in *bazars* in cities such as Lahore or Amritsar, be it obtained from a *Gujar* or a *Ghosi* milk producer or purchased from a shop of a *halwai* in following words:

> The urban milk supply is at present in the hands of *Gujars* in the north of the Punjab, and of *Ghosis* and *Ahirs* in the South...The cowsheds, whether inside the city or in its immediate vicinity, are equally dirty and objectionable. Little attention is paid to the quality of fodder given to the animals. The *Gujar* or *Ghosi* cares only for the quantity of milk he can extract...Milk is hawked around the street in open and unclean vessels, and the sale of milk by *halwai* is conducted in no better fashion...No attention is paid to the cleanliness of his shop. Many remedies have been proposed for this unsatisfactory state of affairs. Sanitary regulations have been enforced and attempts have been made to exclude the *Gujar* from

> the city. *But wholesale condemnation of the present system, or rather want of system, is useless unless an efficient substitute can be provided. Until the people can be shown that a sanitary milk supply is not only possible but profitable both to the retailer and consumer, any measures directed against the present system are useless.* (emphasis added)

However, this report presented the problem as affecting only "a middle-class clerk or small shopkeeper, who is not rich enough to arrange for his own supply of milk, or powerful enough to insist on the improvement of what the *Gujar* and *Halwai* are pleased to purvey to him".[5] The report found that wealthier members of the community were not interested in addressing this problem of supply of pure milk to the public since they "either kept cows in their houses or have a special arrangement with the zamindars of a village outside the city" just like what EBJ and Isa Tweed had imagined would be a strategy employed by a European officer serving in India.

This picture of unsanitary milk supply in cities was presented to readers in narratives penned by elite Indians also, who alongside attempted to conjure up the distant mythical past when Aryans lived in a landscape blessed with plentitude of milk and *ghee* (clarified butter). References to the mythical figure *Kamdhenu* (one of the treasure troves that were attained during the churning of the great ocean by *Devas*—i.e. Gods and *Danavas*—i.e. Demons) or cows kept in the hermitage of great sages—Surabhi, Nandini—were frequently invoked. Partha Chatterjee identifies the tropes of plentitude of nutrition in literary texts by Rajnarayan Basu and his contemporaries, the representative example being *Se Kal ar e Kal* (Those Days and These Days) which got published in 1873.[6] It was not unusual to find books that engaged with the emerging narratives that talked about milk as 'the complete food' and talked about its nutritious qualities as discovered by public health researchers also quoting profusely from Vedic texts and a few treatises on Ayurveda.[7] This tendency to invoke Vedic references and citations from Ayurveda while talking about milk and milk products was so prevalent in India that even an official document, *Report on the Marketing of Ghee and other Milk Products in India* (1947: vii) noted in the preface: "To the vegetarian in India, *ghee* is perhaps the most important article of food. Its importance has been recognized even in the ancient Vedic text where it has been mentioned as "life" itself".

Around 1880s alarmed from narratives on the diseases due to the contamination of milk, municipal towns had started to look for chemists who can be assigned the job of examining the samples of milk and enquire into chemical composition of milk. F. Strohner carried out the initial groundbreaking work by carrying out examination of buffalo milk in Bombay and pressed the Bombay municipality to establish a full-fledged laboratory to continue with the chemical examination of milk samples in 1903. It was at this laboratory that Lemuel Lucas Joshi worked under the leadership of John A. Turner diligently carrying out chemical examination of hundreds of samples of milk over the years. It was the result of such a persistent work by Joshi and other analysts employed at municipal laboratories in cities such a Bombay, Poona, Calcutta, Madras and Bangalore that slowly but steadily the concept of arriving at local standards for milk began to emerge.[8]

In an issue of *Calcutta Medical Journal* of 1910, Rai Bahadur Chunilal Bose had contributed an article dealing with the legislative control that was felt necessary in order to prevent food adulteration.[9] He had followed this essay replete with a tone of advocacy for legislative route with another essay that dealt with the prevailing state of Milk Supply of Calcutta. All the dangers due to 'dishonest *gowala*' and unsanitary conditions of milk being sold in Calcutta city were remarked by narrating a story about how an ancient Indian king set out to test the honesty of (*Aheers* and *Gowala*) milkmen living in his capital: he issued an order that on a certain night every *gowala* should pour in a pitcher of pure milk which he required for a ritual to be performed next morning. It was so arranged that every *gowala* could pour in his share quite anonymously through pipes laid down, especially for this purpose that would lead all milk to flow into a big vessel kept in a locked room so that no one could see, touch or pollute the milk. In the morning, the king had to face a situation where his faith got a rude shock, since all milk vendors "with the characteristic mentality (of their) caste" had presumed it safe to pour in water, rather than milk and the vessel didn't show even a drop of milk. Just before narrating this story, Chunilal Bose had stated, without any apparent signs of self-doubt that "The trade honesty of *gowala* has never been of a high order" and reported that "some of them openly declare that they would be infringing on their caste-rules if they would sell milk without admixture of water, however small the quantity may be".

This essay had talked about an inquiry carried out in 1913 which revealed that 83.6% of the milk supplied to hospitals in Calcutta was adulterated, but in 1916 only 37.6% samples were found to be adulterated and in the year 1917 all the samples were found to be of good quality. However, not much had changed on the quality of milk supplied to the general population of Calcutta. Presenting the data that Bose had obtained from Dr. S. B. Ghosh, Chief Analyst to Calcutta Municipal Corporation, it was reported that between 1913 and 1917, the percentage of adulterated sampled ranged from 26 to 50%.[10]

Thus, we witness that at the Third All India Sanitary Conference held at Lucknow in January 1914; the question of legislating milk standards came up for discussions. The minimum standard of purity in England during that period was 3% of fat content and a feeling prevailed amongst municipal analysts that the English standard was not applicable in India. Presenting his findings of the analysis of 614 milk samples from Bombay, Lemuel Joshi reported that four-fifths of the milk supplied to Bombay was adulterated with water and argued that milk standards for India are essential and a bacterial milk standard was desirable. However, he cautioned that this should not be legislated before more work is done on the bacteriology of milk in India and suggested a tentative standard as to guide administrative action.[11] Four years later, when Joshi published his book, *Milk Problem in Indian Cities* (1916: 04), he reported that "after analyzing 1400 samples of milk, it has been shown that four-fifths of milk supply of Bombay is adulterated with water and over 90% of milk samples examined bacteriologically are contaminated with microbes whose presence indicates dirt etc.".

An essay that Chunilal Bose authored along with Sasi Bhushan Ghose, suggested a minimum standard of purity for cow's milk to contain at least 4% of fat and 8.5% of solids-not-fat, whereas for buffalo's milk the minimum was suggested to be at least 6.5% of fat and 9.5% of Solids Not Fat.[12] Similarly, Lemuel Joshi (1916: 44) outlines the one uniform milk standard proposed to be legalized for Bombay and other cities in western India as follows:

> Milk, as is ordinarily obtained in the market, shall be considered genuine if it is found to contain at least five percent of fat and not less than eight and a half percent of non-fatty solids, provided that the percent of fat may be allowed upto three and a half, only when it is proved to satisfaction of the court that the sample was of unmixed cows' milk, the onus of the proof resting with the vendor.

Around the same time in official circles, there was a thinking that a lot more was required to be done by the imperial and provincial Governments on the subject of cattle breeding and dairying. In Punjab, the Government had decided in August 1909 to publish a report dealing with the cattle trade of the province with special reference to the profits of cattle breeding. The Director of Agriculture was deputed to obtain necessary information on the subject from district officers for this report. This endeavour had resulted in a report titled 'Cattle and Dairying in Punjab' in the year 1910. While the meeting of the Board of Agriculture in India which took place in 1913 at Coimbatore extensively discussed this subject, in the same year, a professional organization—the Dairy Education Association of India—started publication of its journal titled 'The Journal of Dairying and Dairy farming in India'.[13]

In 1920s, protest and boycott actions resurfaced and allegations that were now made against *banaspati ghee* once again reminded about the great fears amongst the *Mahajan* and *Bhadralok* that ghee being sold in Calcutta bazars was adulterated with animal fat that had surfaced earlier in 1886.[14] When Dutch firms began to supply Indian markets with *banaspati ghee* in 1917, this development was looked at as a potential threat that could have far-reaching consequences for native traders who were involved in the trade of *ghee* (clarified butter). A booklet published in 1925 in Gujarati titled '*Vegetable Ghee ane Prani Raksha*' talks about beginning of the import of *banaspati ghee* in Bombay in 1923 and talks about several protest meetings organized by Jain community in Mumbai.[15] The author of this booklet, Nandlal Chhaganlal Adhiya employed tropes of vegetarianism, religious purity, nationalism and *PraNidayaa* (compassion for animals) to argue out that the spread of *banaspati ghee* was a great threat for the animal husbandry in Kathiyawar.[16] He had also hoped to get his arguments against *banaspati* translated into languages spoken in other provinces in India but was sad that there wasn't much progress in that direction and in the meanwhile, there were signs of increase in *banaspati ghee* trade. Referring to data quoted in the debate in the Council of States on 27th February 1929, Mishra (2015: 112) states that "the annual import of *banaspati ghee* rose from nil to the figure of nearly Rs. 20 million". Nandlal Adhiya claimed in his booklet that Indore and Porbandar states had already passed prohibitory orders against *banaspati ghee* and appealed to other native States to follow this decision. Talking about efforts by Jain traders, this booklet refers to the first protest meeting organized by *Kachchhi Visha Oshwaal* youth group (a trading caste community group), the second protest meeting in Kandivali and a third one by *Kachchhi Dasha Oshwaal* youth group in Bombay.

In 1930s, cooperative societies dealing with *ghee* mushroomed all over India and the highest number of them was found in United Provinces. The Ahmedabad

Cooperative *Ghee* Supply Society was established in January 1943, when consumers and *ghee* dealers came together to supply adequate *ghee* of good quality to members. The number of such societies in United Provinces was 392 in 1938–'39, 607 in 1939–'40, 693 in 1940–'41, 700 in 1941–'42 and 727 in 1942–'43. Compared to such swelling numbers of *ghee* cooperatives in United Provinces, the cooperative marketing of *ghee* as an idea had not found much purchase in other provinces. So, in 1942–'43, Madras reported 7 *ghee* cooperatives, Punjab accounted for 10, while Bombay had only 2 of them!

Similarly, the question of adulteration of milk and milk products was on the agenda once again at the first meeting of Animal Husbandry wing of Board of Agriculture and Animal Husbandry in India, held in February 1933. After discussion on this subject, it was recommended that "for the protection of health of people and for development of dairy industry, all legislations related to prevention of adulteration of dairy products should be made obligatory, and not optional, on local bodies". A resolution, to appoint a special subcommittee to work on this aspect and examine all related questions, was passed and following this resolution Imperial Council of Agricultural Research appointed Dairy Legislation Committee.

In the year 1937, the constant articulation of concerns about unhygienic milk and milk product supply had reached such proportion that the government enacted a legislation providing for the grading and marking of *ghee*. The legislation was named The Agricultural Produce (Grading and Marking) Act, 1937. We are told that the enactment of this legislation had "considerably helped the development of the *ghee* industry", which noticed "as much as 48,860 mds of graded *ghee* valued at Rs. 24 lakhs" being put on the market under the AGMARK seal[17].

5.1 Emergence of Kheda as a Premier Dairying District in Bombay Presidency (1850–1940)

After 1850s, Kheda district registered significant advances in dairy industry. In her article that looks at 'Anand model' in the backdrop of the historical roots of dairying, Sujata Patel (1990: 29) refers to a census of cattle that shows an impressive 36.4% increase in the number of buffaloes in Kheda district.[18] James Campbell had shown in the Imperial Gazetteer of Bombay Presidency (Vol III on Kaira and Panchmahal) that as per revenue returns for the year 1876–'77, Kaira reported 1,80,223 female buffaloes against 10,315 male buffaloes; 1,28,247 oxen against 49,261 cows; 56,167 ploughs and 28,971 carts.[19] Amongst other animals, the revenue returns had mentioned 2366 horses, 33,800 sheep and goats and 6235 asses. A careful look at the number of plough and cart oxen and milch buffaloes would make one understand that peasants in Kheda had taken to milk production that went beyond mere household and subsistence needs. Sujata Patel reports that every well-to-do cultivator in Charotar had at least 3–10 milch animals and refers to another report published in 1895 that observed 28% increase in milch animals in Nadiad taluka between 1878 and 1893. Several other academic writings have also tried to estimate the size of milk trade in Kheda district in 1870s. For example, David Hardiman mentioned that during the year 1878, the export trade in milk products from Charotar area was worth Rs. 8 lakh.[20] Talking about the boost in milk economy of Kheda district, Hardiman (1981: 262) states that "The advent of railways and cream separating machines allowed the milk to be sold to the cities either as fresh (liquid milk) or as butter". Thus, by 1918, the dairy exports from Anand town of Kheda district were worth twenty-four lakh rupees a year, which was a sum greater than the total land revenue of the district and suggested a three-fold increase compared to what prevailed some 40 years ago.[21]

The Imperial Gazetteer of the Bombay Presidency (Third volume on Kheda and Panch Mahals) cites a government report from the year 1924, which mentioned that cultivators from Petlad (renamed Borsad later) used to export larger quantities of *ghee*. In their own bullock cart, these cultivators used to make four to five trips to South Gujarat and brought cotton seed back with them in return trip. Imperial gazetteer also reported that not only did the cultivators of Kheda supplied clarified butter to market on a weekly basis, but the district was also seen sending its milch buffaloes by rail in considerable numbers, chiefly to Surat and Bombay to populate city milch stables.

A closer attention to the rural life in Charotar from the memoirs of Bhailal Patel,[22] who served the British State as Civil Engineer in charge of building Sukkur Barrage in Sindh province in 1930s, also tells us that not only this tract energetically embraced cultivating cash crops such as tobacco and cotton, but even amongst food crops, it preferred *bajra*, a kind of millet, which as Russell stated in his review, served dual purpose, "the grain is used for human food, and stalks formed the chief food of the cattle for some six months in the year".[23] Bhailal Patel (1996: 12), a prominent leader from Kheda district, who was responsible for the foundation of the first rural university in Anand town named after the first Home Minister of independent India, Sardar Vallabhbhai Patel, recollects the milk economy of his native Sojitra village in Kheda district during 1888–1900 in his memoirs and states:

> Those cultivators who owned buffaloes got the task to make butter using hand operated wooden churners once in three days, but no one in the village could think of selling the buttermilk. Those neighbours who didn't own buffaloes used to come and collect buttermilk from such households. The poor families, occupation caste families and *Harijans* used to get buttermilk for free. No one could entertain the thought of selling buttermilk in his wildest dream. Every day in evening we had *karhi* made up from buttermilk in our dinner.[24] (my translation)

By the year 1880–81, the Kheda district had very clearly emerged as a region where milch cattle outnumbered plough cattle, cow to female buffalo ratio was more than 1:3. However, the severe droughts and famines in the last decade of the nineteenth century and the severest famine known as *Chhappaniyo*, in the year 1899–1900, wherein 1.3 million people died in Gujarat, seems to have wiped out a large number of plough as well as milch cattle, with young stock being hit hardest. For the number of cattle in different years as reported in the Imperial Gazetteer of Kheda District and in other contemporary sources see Table 5.1.

Table 5.1 Number of Cattle in Kheda district (1880–1915)

	1880–'81	1890–'01	1901–'02	1902–'03	1905–'06	1909–'10	1915–'16
Plough cattle		1,12,871	80,613	84,517	85,010	1,00,906	1,03,544
Breeding bulls		624	268	391	908	947	
Milch cattle	2,02,105	1,55,258	61,602	72,901	67,487	95,958	1,21,853
Young stock		1,14,749	39,899	53,259	68,490	1,00,864	1,62,248
Total cattle	3,52,962	3,84,540	1,83,288	2,11,863	2,22,451	2,99,440	4,19,723
Ploughs	5803	63,122	45,613	47,438	42,865	48,641	57,697
carts	27,719	28,379	23,204	23,465	9482	10,023	24,082

Source Revenue reports and Imperial Gazetteers on Kheda (1878), Bombay Presidency (1903)

Table 5.2 Subdivision wide number of cattle and buffaloes in Kheda

Sub-Division	Oxen	Cows	Buffaloes
Kapadvanj	20,570	14,496	17,249
Thasra	17,572	13,071	15,887
Mehmadabad	14,972	5778	20,729
Nadiad	22,009	3938	30,925
Matar	14,760	5789	19,391
Anand	20,696	2917	41,222
Borsad	17,668	3275	45,635
Total	1,28,247	49,264	1,91,038

Source Imperial Gazetteer on Bombay Presidency, Vol III, Kaira and Panch Mahals

At the subdivision level for Kheda, the imperial gazetteer reports the following figures on oxen, cows and female buffaloes. It also emerges from these figures that within Anand and Borsad (erstwhile Petlad), the trend of buffaloisation was more pronounced (Table 5.2).

Describing the fall in the number of cattle and buffaloes during the last decade of nineteenth century and then a quick bounce back, Crispin Bates (1981: 806) reports:

> Recovering from the years of crisis, farmers in Kheda, Ahmedabad and Broach acquired fresh stocks of high quality Kankreji bullocks from Kathiawar and Sind, and by 1916 there were more plough cattle relative to population in some areas than there were in 1892, recovery being most rapid after 1905. The even more rapid growth in the number of ploughs in the Kheda district suggests that lesser Patidars were now commonly sharing their cattle during the cultivation season (partly a consequence of the continued fragmentation of the larger holdings). *There was furthermore a rapid increase in the stock of high milk yield buffaloes as Charotar Patidars took a commanding interest in the dairy industry, until by 1930s a survey by J. M. Mehta revealed that cultivators in Borsad taluka got more than 70 percent of their subsidiary income from dairying. This was the famous 'white revolution,' doubtless the cause of increased demand for fodder, which began with the establishment of factories for sterilizing and condensing milk at the turn of the century and was considerably encouraged by the First World War.*[25] (emphasis added)

Similarly writing within a few years after the great famine, Joslen (1905: 22) stated that buffalo cows are kept and exported in such large numbers by agriculturists and townspeople in the Ahmedabad and Kheda districts that their maintenance, in addition to supplying their own wants, is evidently a source of profit.[26] Joslen was of the view that for dairy purposes, female buffaloes almost entirely take the place of cows, as they give more milk and manure. He reports that large quantities of *ghee*, butter and cream were exported to Bombay and Calcutta from Kheda district. So, he proposed that Kheda district especially might almost be called one of Bombay's dairies. Joslen reported that buffaloes were being bred in numbers beyond requirements and the surplus, in adult age, were being exported to Bombay and other places. He had also sourced statistics on the amount of milk, cream and condensed milk (*mawa*) being sent away from Anand railway station per week. According to this statistics, cream (desiccated) was the main commodity with 160 cans varying in size that can contain 14 seers to 28 seers (thus on an average 21 seers or 42 lbs.) being booked per week, compared to 139 baskets, each containing 30 seers or 60 lbs. of condensed milk (*mawa*) and 29 cans of fluid milk, each containing 15 seers or 30 lbs. Joslen (1905: 59) had also obtained statistics on the amount of cream, butter and *ghee* as well as cattle booked during the year 1903 from Nadiad railway station destined to reach Bombay mainly. As per this information, 781 mds of cream, 839 mds of butter and 2920 mds of *ghee* as well as 3374 cattle were supplied to Bombay in the year 1903. Joslen's account also makes it very clear that Ahmedabad and Kheda districts used to supply the high yielding breeds of cattle and buffalo for Military Dairy Farms. Around 1915, when several military creameries were established, one of them came up at Anand, to supply bulk butter in 35 lbs tins to troops. During the First World War, about 10,000 lbs. of milk was being supplied daily to hospitals for the wounded and the Government Central Creamery at Ahmedabad was supplying 15,000 lbs of butter daily to the forces at war overseas.[27]

5.2 Milk Boycotts and Peasant Nationalism in Kheda

How did the nationalist peasants of Kheda respond to such a drain of dairy commodities from Kheda and Ahmedabad district in support of the empire's war? Hardiman (1981: 123–125) shows that in 1917, Indian National Congress gave a call for boycott of milk trade in Kheda with an intention to hinder the war effort of the empire, since much of the milk from Kheda trade was being turned into butter for the use of the army:

> It (the milk boycott) was supported by local religious leaders who were alarmed that the peasants were selling, rather than consuming, the elixir of the Hindu life, the milk of the cow. Gandhi gave his support to this boycott for economic reasons. He argued that the insatiable demand for milk from outside Gujarat was raising the price within Gujarat, so that people could no longer afford to drink milk or eat *dahi* and *ghee*... The milk factory owners first began to notice a decrease in milk supplies in mid-1918. By 1919, the agitation had spread to Baroda villages of the Charotar, largely, it was said because Baroda officials

> encouraged it. By 1920, no milk was coming out of the Baroda villages. The movement reached its height in 1921, when two of the leading milk factories reported that they were getting only a quarter of their normal supplies. Milk supply continued at a low ebb in 1922 and 1923, and with the trade depression adding to their troubles, some of the milk factories, were forced to close down. After this the agitation died down.[28]

5.3 The Curious Story of Ups and Downs of Government Research Creamery at Anand

A decade after its establishment, with the creation of the Imperial Dairy Research Institute at Bangalore, the Government creamery at Anand was handed over to Imperial Dairy Expert, Mr. W. Smith in October 1925 to be used as a dairy factory school for study and research purposes. However, even before the much-needed research work could begin at Military Creamery at Anand, the government was again considering to close it down, citing reasons that it had not proved commercially viable. While the Imperial Dairy Expert conceded that "owing to severe competition in the fresh butter market and for other reasons, the working of this creamery had not been as profitable commercially as was hoped", he addressed a letter to the Agricultural Adviser of India on 16th June 1926, registering his protests to this move and termed the closing down of this creamery 'a national calamity'.[29] In his submission to the Royal Commission on Agriculture, he had reiterated his opinion that the research work on the most pressing problems "can only be carried out at an experimental dairy factory situated in the premier milk producing area, Gujarat". He found it useless to attempt this class of experiments with milk produced by Government cattle fed, housed, and looked after under perfect conditions at the Government owned farms at Bangalore and Karnal. He believed that such a research and experimentation work must be done on a factory scale with purchased village milk, produced under ordinary working conditions and adulterated, as the village producer will insist on adulterating it. It was at this site that two young dairy technocrats carried out experiments in the early 1950s to innovate a process of making milk powder from buffalo milk.

It appears that following his very strong protest through the above-cited correspondence, this question had kept him anxious for months. When chairman of the Royal Commission on Agriculture (1926: 19) took note of memorandum submitted by him and followed that with a question, "You are very anxious to keep that creamery in your hand?", the answer that Smith gave was just three words in an assertive affirmation: "Yes, I am". When another Commission member asked him, "Do you not think it would be better to take the opportunity to start a new model factory instead of retaining Anand?" Smith makes it known how dear a place Anand creamery held for him in the year 1926, by stating:

> No. The Military dairies took two years to get a suitable site in days when Government of India had control of the land there. Now, land and all that sort of thing is controlled by the Provincial Government. I think it would be impossible for us to get any site anything like as suitable as Anand for the establishment of an experimental and educational dairy factory. Anand is situated at the junction of four railway lines, right against the station; it has got

> a first-class water supply, and that is very valuable. It is very difficult to get a good water supply for a creamery; a creamery requires a great deal of water.

Answering MacKena, Smith reiterated his plea that Royal Commission should make an interim recommendation apart from their final recommendations, stating: "I have suggested that the Commission should send someone to see the district and see its extraordinary possibilities from a dairying point of view. *It is the one really dairying district in India*" (emphasis supplied).

During the years 1929–'30 and 1930–'31, the Research Creamery at Anand carried out investigations regarding long-distance transport of milk for city supply. Experiments were also tried out regarding the manufacture of dairy products. However, there was a feeling that there wasn't adequate expert staff to enable the work on these lines. Official reports pointed out that "At the Anand creamery, work of a scientific nature was much hampered through the want of new machinery and buildings, particularly a milk-drying plant and a students' hostel".[30]

The proposal for establishing a research creamery for experimental work on handling milk and dairy products at the Imperial Dairy Institute, Bangalore was moved much later by the Imperial Animal Husbandry Expert to Imperial Council of Agriculture Research in the year 1933, the Provincial Economic Conference, 1934 agreed that much-increased facilities were needed for the development of a Dairy Industry in India and as a result Government of India provided a grant of Rs. 6 lakh spread over five years for the same.[31] It is important to underline that this grant also contained a share for improvement and expansion of experimental creamery for experimental work on the handling and processing of milk and dairy products at Anand. The grant also had envisaged to extend the existing facilities at the Imperial Dairy Institute, Bangalore.[32]

However, as we can see from the report of Dr. N. C. Wright in 1937, even this grant of Rs. 1 lakh was lying unutilised and in that year the government research creamery at Anand was again lying abandoned, despite the forceful pleadings of Imperial Expert on Animal Husbandry, Col. Arthur Olver to appoint a dairy technocrat at Anand, with a promise of making him head of the Dairy Technology and Dairy Administration wing at the national research institute when the same comes up in North India.[33]

Following upon these developments, in the year 1938–'39, a scheme for the establishment of Animal Husbandry and Dairying Institute at Anand in Kaira district was prepared and submitted by the Livestock Expert to Government of Bombay. Alas, that proposal didn't receive a positive response from the Bombay Government.

5.4 Imagination on Milk Cooperative Societies in Anand Circa 1928

By the year 1928, Bombay Presidency had about a dozen cattle breeding cooperative societies working in Dharwar and Belgaum while there were three such societies in Thana. In his oral evidence before Royal Commission on Agriculture, while speaking about these cooperatives, Mr G. F. S. Collins, the then registrar of Cooperative Societies had identified those societies in Dharwar as working rather well and those in Thana as not working quite well. In an answer to a very pointed question, viz. whether there existed any urgent demand for (cooperative) societies for cattle breeding and dairy farming, he had stated that unfortunately there wasn't such a spontaneous demand from the people.

The last question which was posed to him does indicate that there was an effort to think of organizing a cooperative milk supply society in Anand way back in 1920s also, may be the one that arose more from the interest of Bombay Municipal Corporation and Department of Cooperation in Bombay Presidency. When we read his oral evidence along with that of Harold H. Mann, we can see that Anand was being imagined as a milk producing centre for the Bombay city. From Harold Mann's oral evidence, we gather that he had proposed that Bombay Government shall extend financial assistance worth Rs. one lakh to such company that was established in a suitable dairying tract with the sole purpose of supplying milk to Bombay city. However, nothing had materialized as the law then didn't permit Bombay Government to assist an entity outside its jurisdiction. That crucial question was: "With regard to the cooperative dairy societies, have you considered what is the minimum limit of finance for a large dairy society, which has to supply milk to a town, for instance?" To this question, Mr G. F. S. Collins replied stating that this question was considered specifically with regard to the establishment of a milk supply society at Anand, and that is all.

Mr. B. S. Patel, the then professor of Agriculture at Agricultural College, Poona indeed was still hopeful of organizing the dairy industry on cooperative lines and had proposed the following:

> At Anand there is a fully equipped creamery of military department, and which is just now being used as a teaching centre and *may be made available for five years for organizing dairy industry on cooperative lines*. What is required is that the Government of Bombay may have to undertake in the first instance to start manufacturing about 1000 lbs of best butter a day and sell it to best market and realise higher price for the best quality. Once this is done, there will be no difficulty in organizing the cooperative milk societies one by one. (Royal Commission on Agriculture, Vol II, Evidence taken in Bombay Presidency. p. 535)

B. S. Patel had also presented these suggestions as a Note that he submitted to Cooperative Conference which took place a little earlier than Royal Commission on Agriculture.

5.5 The Untold Story of Organizing City Milk Supply Through Cooperative Unions

The popular narratives on the milk cooperatives in India have often missed to take into account, the inspirational presence of the milk producers' cooperative societies and cooperative milk supply unions that had started to emerge during the early decades in twentieth century. These narratives that started to emerge during the Operation Flood years (i.e. 1980s), portrays the spectacular rebellion of milk producers against Polsons in villages around Anand in mid-1940s as 'the myth of origin' of cooperative mode of ensuring milk supply to cities. In such popular narratives, the question 'what existed prior to Anand Pattern Milk Cooperatives' gets answered in a very simplistic way by overemphasizing on the 'monopolistic' presence of a private dairy firm in Kheda, Polsons. Such simplification does grave injustice to the history of attempts to organize city milk supply on cooperative lines for two reasons.

First, such simplification fails to account for several proposals that were articulated right from 1920s to organize milk supply to Bombay city by establishing and encouraging milk cooperative societies in Kaira district, the premiere dairying tract of the Bombay Presidency. Second, such simplification also fails to account for pre-independent efforts by nationalist leaders elsewhere in India to organize the city milk supply on cooperative lines, such as Calcutta Milk Supply Union and Madras Milk Supply Union. Those efforts in organizing city milk supply on cooperative lines had resulted in the emergence of a three-tier structure, which seems to have inspired the policy proposed in a report of the Cooperative Planning Committee in 1946. In this section, it is argued that without acknowledging those intellectual debts, Varghese Kurien had championed the same policy under a new name 'Anand Pattern Cooperatives and three-tier structure' during Operation Flood.

The Royal Commission on Agriculture (1928: 219) mentions Telinkheri near Nagpur in Central Provinces, where the local milksellers (gowalis) had been formed into a successful and long-established cooperative society to improve both the feeding and breeding of their buffaloes and cows. It seems that this cooperative at start had just a dozen gowalis who came together and formed a milk supply society in a suburb in Nagpur city and Government had also started a Dairy Farm at Telinkheri village on the outskirts of Nagpur city in the same year. When neither gowalis nor Government Dairy Farm could earn profit, they decided to combine their operations. After gowalis' herd consisting of cows and she-buffaloes was added to Government Dairy Farm, the average yield of combined herd reached a level of 800 lbs per day. Five years later, Gowalis of another suburb also joined the society.

This cooperative had nearly 200 milch buffaloes and cows that were kept along with Government herd at Telinkheri and were crossed with Government bulls, reported D. Clouston, the then Deputy Director of Agriculture in Central Provinces in a note written for ninth meeting of Board of Agriculture. Clouston (1916: 133) stated, "fine Delhi milch buffaloes are being bred and crosses are being produced between the Delhi and the local breed". Clouston, who had referred to gaolis in the same note as "illiterate breeders" whose methods of breeding and feeding were "of the most primitive kind" went on to record in the note as if in a self-congratulatory style, "this is an honest endeavour to show our professional milkmen how to improve their lot". In 1913, another cooperative dairy, Katra Cooperative, Allahabad was also registered. This cooperative dairy also worked successfully for many years and "during 1936–'37 it had one primary society of 45 members and was handling on an average about eight maunds of milk daily".[34]

While the idea of organizing milk producers' cooperative societies—and in a similar style—cattle breeders' cooperative societies had not found a widespread acceptance by 1920s, the thought that this mode of milk production and supply was the surest way to address the problem of unsanitary situation in city milk supply found repeated articulations in official discussions. Thus, at the ninth meeting of the Board of Agriculture in India held at Pusa (Bihar) in 1916, S. S. Milligan voiced his belief that "the distribution of pedigree bulls, the establishment of milk records, the spread of knowledge amongst the people about the organization of dairy concerns and the marketing of dairy produce can be best done through cooperative societies".

The most promising cooperative movement was witnessed in the form of Calcutta Milk Supply Union, which aimed partly to improve the city milk supply and partly to enable cultivators in neighbouring villages to break the middlemen's cartel, and thereby secure a fair return for their produce in the year 1915 under the guidance of Mr. Donovan.[35] The success of this society had a ripple effect amongst nearby villagers and within three years 54 cooperative milk societies—with an average membership of 45 milk producers—got affiliated to Calcutta Milk Supply Union. A decade later, the milk supply came from affiliated village societies that numbered 71 in 1925–'26, with membership running to 3600 members. In five years, the membership had gone up 50%, but the milk supply had trebled, for the average quantity per member had risen from 0.56 to 1.15 seers (1 seer equaled to 1.25 kg or 2.75 lbs) per day. The village societies collected milk and sent it to the nearest depot and there were six of them, from where the milk was transported by rail in sterilized cans to Calcutta. The village societies sent in approximately 850 gallons of milk daily between February and July and 500 gallons between August and January.[36] During the year 1926–'27, the number of cooperative milk societies in Bengal increased from 76 to 97 and 82 of these societies were affiliated to the Cooperative Milk Union in Calcutta. During this year, Calcutta Milk Union had booked a profit worth Rs. 12,148 on the sales worth Rs. 1,54,122. Such a spectacular show had resulted in affiliated milk cooperative societies registering a reserve fund worth Rs. 35,844 as against the share capital of Rs. 10,069.[37]

Reporting on the organizational aspects and working culture of the cooperatives, Royal Commission on Agriculture (1928: 66–67) states:

> Some reference to the work of Milk Union, Calcutta and its dependent societies may be of interest in view of the importance of obtaining for Calcutta an adequate and pure milk supply. The operation of each of the 71 milk societies constituting union is generally confined to one village, the members of which are *bona fide* milk producers whose primary occupation is agriculture with milk production taking a secondary place. The societies are arranged in groups for the purpose of collecting milk by paid milkers and carriers. Each group consists of six or more societies. The milk obtained from the societies in a group is collected at a depot,

> which is under the charge of a depot manager, whose duty is to receive the milk in properly sterilized cans, measure it, note the general conditions by lactometer and give a receipt to the carrier. The work of the depot is looked after by a depot supervisor who is maintained by the union. The milk when received by the union is distributed by their staff, either salaried or paid on commission, or by traders who purchase their supply outright. The union has established distributing centres at convenient places all over the country. Government assists the movement by lending a veterinary assistant surgeon who examines and treats the cattle belonging to the societies, supervises the market arrangements and looks after the sanitary conditions of the cow sheds.

Reporting on the financial side of this union, Royal Commission on Agriculture (1928: 67) notes:

> The operations of the union are at present on a very small scale; its sales on average throughout the year are only about 70 *mds* of milk as against an estimated daily consumption of 3000 *mds*, in the city of Calcutta. The experiment is nevertheless, an important and hopeful one. The financial position of the union is sound. Every year, except in the year of its foundation (1919-'20), it has made a profit; the profit has steadily risen and out of total working capital of Rs 75,368, only Rs 15,834 or 20 percent has been obtained from outside source.[38] The way in which the union spends part of its profit is of striking interest. Rs 800 goes towards the education of sons of the members of the societies by means of assistance given to five primary schools and one secondary school in the area in which the union operates. In addition, it has given a donation of Rs 600 towards the construction of two primary school buildings. The union has also arranged for the delivery of the lantern lectures on sanitation, anti-malaria and anti-*kalazar* work, and has made provision for treating any of its members who may be suffering from malaria or *kalazar*. Finally, as a work more strictly relevant to its object, the milk union has distributed fifteen stud bulls to the primary milk societies and has made provision for replacing them when necessary.

P. J. Kerr, Veterinary Adviser to the Government of Bengal Presidency notes in his memorandum submitted to Royal Commission on Agriculture in India (1928: 98) that "the Cooperative Milk Union at Calcutta has been successful in inducing cattle owners to take a keen interest in their cattle, and so be willing to accept advice as to their improvement by better keeping, feeding and breeding". Kerr held an opinion that "the formation of such milk societies in all places where there is a ready market for milk, or its products would be of great incentive to improved methods of animal husbandry". Kerr stated that although the union had begun work with "the main objective to solve the pure milk supply for Calcutta," he had followed its work from his "official interest from the point of view of the improvement of cattle". Reporting on the change induced by the union, Kerr reported that "the members of these milk societies have been willing to accept advice on the feeding of their cattle, the purchase of better cows and bulls, and lately have evinced a desire to have their cattle 'permanently immunized' against rinderpest, at a cost[39] of Rs. 1-8-0 to Rs. 2-0-0 per head". Kerr also notes that the societies had also contributed to popularizing the cattle show held at Baraset, and in a few years became successful to transform it from that attended by a few persons to quite a large show attended by two hundred or more. Kerr had desired to replicate this model with a view to "exploit milk production in Bengal to the utmost, and where it is not possible to sell milk, to investigate the possibility of establishing factories for milk products".

During cross-examination, the Royal Commission on Agriculture members asked Kerr, "is it (milk union) capable of expansion?" to which Kerr's reply was very much in affirmative and he went on to suggest that while at that point of time, the union was bringing in around 120 mds of milk per day, they were "under an obligation to the Corporation to increase that to 200 mds in 12 months and to 400 mds in four years from the time they got the loan". Rao Bahadur J. M. Mitra, the registrar of Cooperatives (1928: 132), also held an opinion that "a multiplication of these societies and their extension to other areas would secure a great improvement in methods of animal husbandry".

The Imperial Dairy Expert, W. Smith, in his oral evidence before the Royal Commission on Agriculture (1927: 22–23) mentions that this cooperative movement had frequently sought his advice ever since he was appointed as Imperial Dairy Expert. He had known the work being done by them in Bengal very intimately and had a very positive opinion about it being a sound and well-managed concern. While in his schema of formative thinking during late 1920s as the Imperial Dairy Expert, W. Smith gave priority to the research work that he was carrying on at Imperial Dairy Research Institute, Bangalore, he was of the opinion that "as soon as we are able to say definitely on how and under what conditions we have to carry out Dairying and Cattle Breeding and get the results of our research work, I should be greatly in favour of getting to the cultivator through the agency of the Cooperative department." What he states, exactly after these words, is something remarkably pro-producers and pro-cooperatives and one would like to cite this formative thinking of W. Smith, keeping in mind that he uttered these words being fully aware that he had witnessed private trade in milk and milk products by the likes of Keventer and Polsons as well:

> I believe that this particular industry, the manufacture of milk products, is especially suited for being carried out on co-operative principles. That was the very reason why I started this class for co-operative officers to get into touch with these men to find out what their ideas were and how much we could assist them. In my opinion, in this country, if industries of this sort are started by private enterprise, the real producer of the milk will not get a proper share of the profit. In practically every country in the world where they have made a success in dairying, they have taken up co-operation for the factory part of the work. It has been particularly suitable.

During the years 1929–'30 and 1930–'31, Calcutta Cooperative Milk Union made further progress in Dacca, Chittagong and Calcutta, under the guidance of the dairy section of the Registrar of Cooperative Societies, Government of Bengal. By now, the Calcutta model of organizing city milk supply on cooperative lines was being tried out elsewhere and in down south, we witness Government of Mysore undertaking a survey of milk supply to Bangalore city and envisioned erecting a modern cooperative milk factory in Bangalore city.[40] Just a couple of years before this, an initiative was tried out to organize milk supply to Madras city by organizing a dozen milk supply societies in Chingleput district in December 1927.[41] Initially, the Madras Cooperative Milk Supply Union had 14 societies and 15 individuals as its members. The quantity of milk sold during the first year amounted 17,408 Madras measures.[42] In the year 1929–'30, Madras Cooperative Milk Supply Union undertook the supply of milk to some of the State Hospitals also. While the supply was quite regular for first few months, there were challenges when certain outside influences from private milk firms engineered disloyalty amongst union members who went on strike for few days and the Union had to pay Rs. 17,000 to the Government as damages.[43] In 1929–'30, the Union made Rs. 1,56,754 from the sale of milk and milk products. A decade later, in the year 1938–'39, Madras Cooperative Milk Supply Union had 14 primary societies with 800 members and handled 88 mds of milk daily on an average.[44]

N. N. Bose, the Superintendent of Calcutta Milk Union, in a note prepared for the Fifteenth meeting of Board of Agriculture (1931: 105) held at Pusa on the 9th of December 1929 and following days argued out that "the steady growth of Cooperative Milk Unions and Societies not only in Calcutta but also in other parts of Bengal has fully established the necessity of organizing and developing the milk industry on cooperative basis". The Calcutta Milk Union which was established in the year 1919–'20, in its more than decade long career then had emerged "from a stage of experiment to that of assured success". The Cooperative Milk Organization in Bengal had a two-tier structure: (1) the rural milk societies which formed the producing centres and (2) the Milk Union in the city which is a financing and distributing agency. Talking about what was aimed with this two-tier cooperative milk organization in Bengal, Bose (1931: 106) states:

> The aim of Co-operative Societies for the production of milk combined with the sale of milk is to get rid of middlemen whose interference in the town accounts for the low price often received by the producers. The problem is to replace them by an efficient democratic organization of the producers themselves, which would give better services than the parasites who ought to be eliminated.

The Calcutta Milk Union, which had 71 affiliated milk producing societies in 1925–'26, had grown into a steadily progressing union with 104 affiliated milk producing societies in the year 1929, with a working capital of Rs. 2,80,000. The milk distributed by it had risen from half a maund in the year 1919 to no less than 150 mds or 12,000 lbs in the year 1929, and it did the business worth over 5 lakhs of rupees.[45] The pasteurization dairy plant that Union had established in Calcutta with automatic bottling, capping and washing machines, according to the plan and specifications by the Imperial Dairy Expert W. Smith, was regarded as most up-to-date milk plant in India. The union had distributed 20 stud bulls and organized 4 cattle breeding societies in its area of operation and encouraged the cultivators in the growing of fodder crops by the free distribution of seeds. Out of its profits, it used to spend liberally for the welfare of milk producers, by putting aside Rs. 800 for the upkeep of cows and dispensary and spent Rs. 1000 annually in aid of schools.

Talking about the ripple effect of this successful example of Calcutta Milk Union in expanding the Cooperative Milk movement in six districts of Bengal and the formation of five milk unions, Bose (1931: 106) stated:

> Out of five Milk Unions, Darjeeling Milk Union is still mainly a Creamery Society and the Shahzadpur (Pabna) is manufacturing *ghi* only. To the best of my knowledge nowhere in India can the milk supply of a single village compete with the milk production of the villages in Darjeeling district. One society alone produces 20 mds of cows' milk which is rich in fat. It varies from 6 percent to 7 percent. The construction of the dairy building of the Darjeeling Union has been taken up during the year, and the factory working on gravitation system

will be first of its kind in India. All the Unions are working at a profit. The possibilities in this direction are indeed vast and immense but the difficulties are equally great and many. In the first place, we cry not so much for money as for men. It is very difficult to find the right type of officer who combines business sagacity and driving power with expert technical knowledge and skill and who really feels for his less fortunate brethren. Secondly, it is to keep the supply regular throughout the year. How to make up the deficiencies in the tight season and how to dispose off, to the best advantage, the surplus in slack season are our greatest problems…. In conclusion, I may remind you that the Cooperative Milk Movement is essentially a movement of actual producers. It gives much more to the producers without increasing the costs to the consumers and, therefore, encourages the producers who deserve a lot of encouragement to continue their hitherto profitless trade with tremendous zeal.

By the year 1935–'36, Calcutta Milk Union was organized into 123 primary societies, with 7500 members, a share capital of Rs. 18,000; a reserve fund of Rs. 72,500 and a working capital of Rs. 99,300.[46] In the same year, Bengal Presidency reported having 247 milk cooperative societies, which were federated into five milk supply unions, namely Calcutta Milk Union, Dacca Milk Union, Chittagong Milk Union, Naogaon Milk Union and Darjeeling Milk Union. By the year 1936–'37, Calcutta Milk Union had 124 affiliated primary societies with 8359 members and handled on an average about 104 mds of milk daily. Giving details of the steady progress of Calcutta Milk Union, during late 1930s, Mukherjee (1937: 230) states that "In 1935, the primary milk societies supplied to the Union milk valued at Rs. 1,49,928 as against Rs. 1,69,000 in the previous year and this decline was mainly due to prevailing economic depression".[47]

Talking about the difference Calcutta Milk Union had brought about in the lives of milk producers, Eleanor M. Hough (1932: 162–163) stated:

Bengal is pre-eminent in the cooperative dairy field…Producer members (affiliated to Calcutta Milk Union) are reported to have improved their economic position considerably, having paid off debts, bought more cattle, erected decent cattle-sheds, and accumulated a substantial sum in owned capital in their local society. Its success has led to the considerable expansion of the movement and has resulted in the creation of a new official post, Superintendent of Milk Societies, the holder of which assists in their promotion…The milk societies, by giving members a better price for their products, amounting in the case of one Bengal Union to 75 percent more than the dairy farmers had been getting, give an incentive for increasing production and so stimulate the breeding of improved strains. Apathy of consumers to the advantages of pure milk is reported as a general obstacle, the removal of which must wait for the education of public opinion.

Eleanor Hough, who had come to India as a doctoral researcher pursuing her Ph. D. from George Washington University, also talks about another model of milk cooperatives with a mixed memberships of producers and consumers and quoted a small cooperative dairy at Surat in Bombay Presidency as an example. She also noted that this cooperative dairy society was "gradually eliminating the latter (i.e. consumers) because of difficulties in recoveries on advances made to them". Quoting the opinion of the Registrar of cooperative societies in Bombay Presidency regarding the working of milk cooperatives in 1930s, Hough stated, "the Registrar points out that they could be more successfully organized if the municipalities would take an interest in the work and give facilities for building cattle-sheds or make loans to the societies at low rate".[48]

On 15th July 1937, the Coimbatore Cooperative Milk Supply Union was established with efforts by Mr. T. L. R. Chandran (ICS), Additional Joint Registrar of Cooperative Societies. The union grew from strength to strength, due to the untiring efforts of persons like C. S. Rathnasambapathy Mudaliar and M. S. Palaniappa Mudaliar. The Union had only seven affiliated milk producers' cooperative societies before 1942, and all of them were within a distance of 7 miles from Coimbatore.[49] In March 1938, a Cooperative Milk Supply Union was established in Lucknow due to great interest shown for the task of organizing city milk supply on cooperative lines by the Congress Government when they were in office in United Provinces. The provincial government extended a helping hand by giving a grant of Rs. 25,000 for the purchase of machinery and motor van.[50]

In his *Report on the Development of the Cattle and Dairy Industries of India* Wright (1937: 21) considered it "necessary to show that the improved methods (of handling and processing milk) are capable of application under village conditions, and that they result in the creation of a better market for the liquid milk". As examples of centres, where this had been attempted with much success, he cites Telinkheri Cooperative Dairy at Nagpur, Cooperative societies at Parbatpur, United Provinces and other neighbouring villages and Singhal's Dairy at Itaura, United Provinces.[51]

By the year 1938–'39, two major Cooperative Milk Supply Unions had established some evidence that the question of city milk supply could be tackled by organizing milk supply on cooperative lines. However, the share of such cooperative milk supply in the total amount of the milk supplied and consumed in city had remained less than 10%: Madras union handled 6.9% the total quantity of the milk market, while Calcutta union handled 1.8% of the total quantity of the milk market. While in percentage terms this quantity may appear miniscule, one also needs to understand that the state patronage was not very high for these cooperatives and the competition from private firms a constant menace to tackle. Despite these challenges, a contemporary account lauded their efforts by stating that "there was no other cooperative organization anywhere else in India which handled milk as much as these two unions".[52]

The Annual Report on Co-operative societies in Bengal for the year ending June 1941 (1942: 27) indicates that Calcutta Milk Union was now organized into 125 primary societies, while there were 223 primary milk producers' societies in entire Bengal Presidency. The report stated that while there existed four milk unions, the Calcutta Milk Union alone continued to do business on a big scale and reported that:

> Calcutta Milk Union sold 43,497 mds of milk for Rs. 3,55,508 against 40,244 mds for Rs. 3,10,388 in the previous year. The total milk handled was 45,254 mds against 41,459 mds in the previous year. The daily average of milk handled was 124 mds and the average wastage was 2 mds 35 sers per hundred maunds, i.e. about 3.5 mds per day. The cost of management was reduced to Rs. 52,572 from Rs. 77,684 in the previous year.

This annual report suggested that despite handling such an impressive volume, the Calcutta Milk Union still faced the problems that were underlined by N. N. Bose in his note prepared for the fifteenth meeting of Board of Agriculture in India. It stated that "the main problem of the Union which so far baffled solution is the high percentage of milk "wastage" associated with which is the question of dealing with the excess supply of milk during one part of the year and shortage during another".

The Annual Report on Co-operative societies in Bengal for the year ending June 1943 (1944: 11) indicates that the Calcutta Milk Societies Union had incurred loss of Rs. 3 lakh in the financial year 1941–'42 and was improving its position a bit by booking a profit of Rs. 14 lakh during the year that followed a famine that must have proved difficult for the members of primary milk cooperative societies affiliated to it. For the year ending June 1944 also, the Calcutta Milk Union retained the profit level at Rs. 14 lakh only.[53] However, *The Annual Report on Co-operative societies in Bengal for the year 1945* (1946: 02) states, "The Calcutta Milk Societies Union earned a profit of Rs. 31 lakh as against Rs. 14 lakhs of the previous year". This report also mentions that besides the Calcutta Milk Societies Union, there existed in Bengal Presidency two other dairy cooperative unions, namely The Chittagong Central Cooperative Milk Supply Union and The Darjeeling Cooperative Creameries Union.[54]

In July–Sept 1948 issue of *Indian Cooperative Review*, a paper on Calcutta Cooperative Milk Societies Union talked about difficulties in the way of expanding business in the following words:

> Included in the category of difficulties is the seasonal variation of the supply of milk by the societies, where only cows are maintained to the exclusion of buffaloes and also absence of any binding obligation on the societies to maintain a reasonable quota of supply and not to divert any part of the cooperative produce to unscrupulous agents and middlemen. In consequence, the Union has to face the problem of having to receive a smaller supply of milk in the dry season and a greater quantity in the flush season than it actually requires. This also necessitates the system of cash purchase of milk at high prices from non-members of the city during the period of September to January to make up the short supply and disposal of surplus milk of societies at abnormally low rates during the period from April to June. Both the systems involve the Union in the losses.[55]

While what happened to Calcutta Milk Union in years after 1948 remains unexplored and unknown,[56] it does remain a very successful initiative from pre-independence years in the arena of cooperative dairying. The two-tier structure that evolved from it, appeared to hold promise of maturing into the three-tier structure, when other districts in Bengal formed similar Cooperative Milk Unions, characterizing a blueprint of 'Anand Pattern' during the Operation Flood years. The way N. N. Bose had expanded these milk unions in the early 1930s in Bengal raises the question of what might have happened to those five milk unions, if those happened to run profitably for more years and combined their milk marketing operations under an umbrella federation. The pattern would have most likely evolved into these five or more unions forming a 'marketing federation' to serve the nooks and corners of Bengal Presidency.

One can indeed trace such an ideation in the policy proposal outlined in the Report of the Cooperative Planning Committee (1946). This report (1946: 46), noted that "before the war, cooperative milk supply societies and unions supplied about 1/8th part (or 12.5%) of the demand of urban fluid milk market, but recently attempts have been made to increase their share in milk distribution in Madras, United Provinces and Bengal". The report also mentioned the fact that as of 1946, there existed 24 milk supply unions and 388 primary milk cooperative societies in British India. The report recommended "formation of cooperative societies of milk producers in the villages situated within a radius of about 30 miles from towns with a population of 30,000 or more" and "especially in those areas lying beyond a distance of 10 miles of the town". It also recommended that a federation of such primary milk cooperative societies would emerge as milk supply unions. The report imagined that such unions would pass on the raw milk collected from primary societies on a contractual basis to consumers' milk distribution societies, which would be responsible for its pasteurization and distribution. However, in the case of large cities, with a population of one lakh and above, the report recommended the establishment of statutory Milk Marketing Boards. The report had however, underlined that in the large cities where cooperative societies for the supply of milk already exist and are functioning well, or can be organized, no Milk Marketing Board need be set up, as those milk cooperative unions can handle marketing.

While we have discussed the milk producers' cooperatives and cooperative milk supply unions at length, let's now talk about how the route of cooperative action was invoked in the sphere of 'cattle breeding' and 'cattle insurance'. Unfortunately, cattle breeding cooperatives could gain a foothold only in Punjab and cattle insurance cooperatives couldn't chart out quite a successful history during pre-independence period. Talking about the cattle breeding cooperatives that existed in late 1920s, Eleanor Hough notes:

> Bengal (has done a little) with cattle breeding and the Bengal Registrar reporting that one or two such societies have shown signs indicating that, if properly run, cattle breeding (societies) is likely to be a success. Ajmer has two active cattle-breeding societies. The development is more extensive in Baroda (which surprisingly didn't have a singly milk producers' cooperative and the Registrar reporting that neither producers nor consumers were interested in establishing one), which at the end of 1928-'29 had five cattle-breeding societies proper besides twenty-nine milch cattle societies which are really credit societies with the specific purpose of financing the purchase of buffaloes of good breed. Here again, Punjab leads with 150 stock-breeding societies and twelve milk-recording societies in 1929–'30.

In 1946, the Report of the Cooperative Planning Committee stated that the number of cattle breeding cooperatives had increased further, and Punjab was still the province that registered great strides in this aspect by reporting 739 cattle breeding societies out of the total 824 such societies in British India.[57]

A review of the history of early milk cooperatives in India and the emergence of divergent modes of city milk supply and the private trade in milk and milk derivatives shows that Kheda had clearly emerged as a tract showing promise for dairy development, right from the first decade of the last century. However, till the late 1940s, the marketing of milk and milk derivatives produced in Kheda continued to be handled by private concerns. This review also raises an important question, regarding the place of Anand based Kheda District Cooperative Milk Producers' Union (KDCMPU) in the country's dairy development history: why is it that a very successful initiative in dairy cooperatives from eastern part of this nation has remained an almost untold and unknown story in popular memory?

Whatever those technical reasons for the strides of KDCMPU are, this narrative reminds us that the history of milk cooperatives pre-dates Anand pattern cooperatives and shows how it is far too simplistic to answer the question: 'What existed prior to Anand Pattern cooperatives?' in just one word: Polsons! While it might have a semblance of truth that Kheda district during pre-KDCMPU times had witnessed private trade, it would be wrong to propose that Polsons had always been a monopolistic presence there. Private trade in Kheda had many players apart from Polsons and Ruth Heredia's book *Amul India Story* talks about those players such as A. C. Stafford of the British Express Dairy,[58] Indian Dairy Supply Company at Nadiad,[59] and Doshi Brothers. The monopolistic presence that Polsons tried to gain in Kheda district had come about only following an executive order of Bombay Government, which led to a spontaneous uprise and formation of milk producers' cooperatives.

5.6 Depicting City Milch Stables as Effecting Serious Drain and Efforts to Relocate City Milch Stables

Riding on the wave of soaring milk prices in cities, there had emerged a model of dairying in late colonial India, where milch stables commanded a huge share of city milk supply. Those who managed these milch stables in cities looked upto milch cows and buffaloes as mere milk producing machines. City milch stables were sites of quite a different mode of milk production where cattle were not integrated with crop cultivation or derived their feed from village grazing grounds. When the milch animals housed in these stables ceased to be in-milk, they were driven out or sent to slaughterhouse. As we can see in several official documents, the milch stables located in the heart of the Bombay city or Calcutta were talked about with a contempt from several perspectives. The cow protection groups felt that these stables were responsible for the deterioration of famed milch breeds since the calves were not allowed to attain maturity and be available for breeding purposes and cows and buffaloes were pushed to slaughterhouses once they went dry. Others considered these stables as an impediment in ensuring proper sanitation and hygiene.

In their 'Note on Dairying in India', prepared for discussion at the Board of Agriculture in India meeting held at Coimbatore in December 1913, S. Milligan and C. M. Hutchinson describe city milch stables as "effecting a serious drain on the best milch cattle in the country". Emphasizing the need for a government intervention to address this serious issue, they stated:

> According to the experience of people who are in the habit of buying large numbers of dairy cows and buffaloes, it is becoming increasingly difficult to obtain good milking animals. Although it is unlikely that within the recent years the general milking capacity of cows has deteriorated from natural causes, there is little doubt but that the town dairies are producing a serious drain on the best milking cattle in the country. The percentage of good milkers amongst India cattle is so low that this may in the end bring about serious results. The question of preservation by government agency of the best milking cows in the country is therefore an urgent one.

When one refers to the discussion that took place at the meeting, one gets a sense that this sentiment that city milch stables effected "a serious drain on the best milch cattle in the country" was echoed by other officials:

> Mr. Wood referred to the heavy loss des to the importation into large towns of the best milking animals which were not subsequently bred from. Mr. Smart and Mr. Hamilton confirmed this from their experience particularly as regards buffaloes. Mr. Smith said this drain was most serious but pointed out that the committee's recommendation dealt with this in general way; if the problem of city milk supply could be solved by the importation of milk from outside and the elimination (sic. probably Smith was suggesting of removal and rehabilitation) of milk producing animals from urban areas, the drain would cease.

At the end of this discussion, the president of the Board of Agriculture in India proposed and Dr Leather seconded the following resolution, which was passed unanimously:

> The Board accepts the general principles enunciated in the report of the Committee on the Dairy Industry in India and desire to emphasise the importance of a satisfactory solution of the problem of the milk supply of towns, not only in itself, but from the point of view of stopping the drain of good milch cows and buffaloes to the towns where they are not utilized for breeding purposes to the best advantage. The Board consider that some legislation is necessary.

However, the first efforts to remove milch animals outside city limits and push the owners of milch stables to designated 'milk colonies' gets effect in Bombay only two decades later in the year 1936–'37. Right from 1913 till 1936, what we come across repeatedly are the sentiments that looked at city milch stables and city-based slaughterhouses jointly producing devastating results by effecting "a serious drain on the best milch cattle in the country".

Report of a Committee on Milk Supply to Towns in Bombay Presidency (1923) notes that "existing conditions under which milch buffaloes are kept in cities tend to the slaughter of the animals who are of greater value for milk production, since they do not get covered when they come in season and in many cases do not come into season owing to unnatural conditions, and therefore are not usually in calf when they go off milk as they ought to be." The committee observed that this was the case particularly in the Bombay city, where it estimated out of total 21,000 buffaloes housed, 7000 buffaloes (i.e. 33%) were getting slaughtered annually. Similarly, in Calcutta data for the year 1913 puts the numbers of cows slaughtered at 9147. Joshi (1916: 09) cites a recommendation put forward in the report by the Special Committee on the Milk Supply in Calcutta that, "the slaughter of prime cows shall be discouraged and the fee for slaughtering such cows should be tentatively raised to Rs. 10."

In a note that Zal R. Kothawala, Dairy Superintendent of Bombay Municipality prepared for the Cattle Conference at Bangalore (1924: 50), he stated that "one of the most unsatisfactory features of the milk supply of cities in India is that the major portion of the supply is obtained from cattle stabled in the very heart of the cities". Kothawala believed that "the first step, in the improvement of the milk supply of any Indian city is the removal of the stables from the towns and the supply of milk from rural areas." This was also a belief that gets reflected in the writings of the Imperial

Dairy Expert, W. Smith as he repeatedly outlined how London and other European cities had stopped housing milch cattle in stables within city limits right from the year 1860. By the mid-1920s, the problem of city milch stables had acquired such proportions that Kothawala (1924: 50) stated that it had "not only drained the Bombay Presidency alone of its best breeds, but its effects are felt even in North India whence the Delhi buffalo is imported in fairly large numbers for the Bombay milch cattle stables". Similarly, C. F. Payne, Chairman of Calcutta Corporation had observed that "the best cattle in their prime are sent to towns, where they are subjected to cruel Phooka process and as soon as they cease to give milk, they are handed over to the butcher".[60] R. C. Wood, Director of Agriculture in Madras Presidency, in his Note on Cattle remarked, "[M]ost of these cows, which come to supply the demands of the Madras milk trade, go to slaughterhouse as soon as they run dry, and calves are allowed to die of neglect".

Kothawala reported that to tackle the problem of "wanton destruction of milch cattle in their prime age, the Bombay Municipality had decided to put a stop to the slaughter of all animals below the age of 8 years". In his note, Kothawala had also referred to a scheme drawn out to rehabilitate the city-based milch stables by providing a set of model stables at Trombay, about 12 miles from the city centre. He also underlined plans put forth by a private entrepreneur to open two dairy farms, one at Talegaon and other at Bulsar, with commitment to supply all the milk produced to Bombay city. However, this scheme of removing the city milch stables had not materialized due to paucity of funds.[61]

Around mid-1930s, the local government, set apart 1054 acres of land at Palghar to be leased at concessional rate for dairying purpose to owners of milch stables willing to relocate away from the city centre and establish a new dairy area for the supply of milk to the city. During the year 1936–'37, two areas of 500 and 537 acres, respectively were allotted to approved applicants and it was hoped that "these grantees work the conditions of the grant satisfactorily and so give a lead and an example to subsequent participants" in the Palghar Dairy District Scheme.[62] However, during the very next year, Annual Report of Department of Agriculture, Bombay Province, noted that some difficulties were being experienced in the working of the scheme, due to the grantees of the land not working up to the conditions of the grant.[63]

In 1946, the Report of Cooperative Planning Committee had outlined what was being thought in official circles on how to salvage dry cows and buffaloes from city milch stables and how to rehabilitate the operation of these private milch stables by organizing them in proposed 'milk colonies' that were modelled after the Military Dairy Farms. Here is the recommendation that was put forth by them:

> The conditions under which buffaloes and cows are stabled in large cities like Bombay are, to put it mildly, appalling, unhygienic, insanitary, unnatural and expensive. It has been suggested that dairy farms, on the lines of military dairy farms, where animals can be maintained on colonies, should be established in the outskirts of towns and cities. These farms should be organised on co-operative lines, and owners of milch cattle particularly 'gowlis' and others owning a few cattle in large cities should be induced to form cooperative societies. They will settle in these colonies with their cattle... Such societies should be offered by Government or local bodies sites on attractive terms, for keeping their cattle, at convenient distance from the cities...These cooperative cattle owners' societies should be able to adopt improved methods

of milk production which have already been current in the military dairy farms. They should also be able to engage some of the personnel which will be released from the military dairies in the course of time. It will also be possible on these farms to arrange for the proper care of dry animals and rearing of the young stock and to provide facilities for grazing, silage making, etc. These societies will be in a position to supply regularly the milk produced by the cattle kept by them to milk unions or consumers' milk distributing societies in the urban area.

There in those words was a blueprint laid out of what was going to be Aarey Milk Colony, dreamt about by Dara N. Khurody.

5.7 Organizing City Milk Supply: Bombay Shows the Way

While a lot has been written about that moment when Polsons was given the monopoly rights for procuring milk for Bombay Milk Scheme by Government of Bombay, here is an excerpt from an official report dating 1944–'45 that suggests under which circumstances Anand, Borsad and Nadiad emerged as promising talukas for organizing a swift milk collection and supply to satiate the thirst for more and more milk in Bombay in mid-1940s.

> During the year 1944-'45, the problem of milk supply to Bombay city had become very acute due to the increase in the population and territorial restriction on the imports of animals in-milk. Government had issued several orders in order to improve the milk supply and to arrange for its proper distribution amongst the population. As a result of these frantic activities, rapid surveys were carried out in Kaira, Panch Mahals, South Gujarat and East Khandesh in order to investigate the possibilities of increasing the milk supply. These surveys revealed that the only tract which has ample milk and good organization to collect it is Anand, Borsad and Nadiad talukas of the Kaira district and arrangement are being made to get daily quantities up to 50,000 lbs of milk to Bombay from the milk producing tract round about Anand.[64]

In June 1945, an organization called People's Provincial Food Council, Bombay published a small booklet titled 'Milk Problem' that was written by J. N. Mankad.

5.8 Kheda District Cooperative Milk Producers Union Arrives (1946)

Report of the Cooperative Societies in Bombay state for the year 1949-'50 highlights the contribution made by KDCMPU to milk supply to Bombay city. The Government of Bombay had identified three agencies in Kaira district at Anand for collection, processing and transportation of milk to Bombay on government account. There were 16 cooperative societies of milk producers in Anand area affiliated to Milk Supply Union at Anand and they had supplied 47,26,718 lbs of milk (i.e. 8,91,833 L of milk) during the year valued at Rs. 10,25,918. This roughly translated into around 2500 L of milk being supplied on behalf of 16 milk cooperatives every day during 1949–'50.

However, the above figures don't reflect the total quantum of milk handled by the Milk Union, since the report also stated that "the union collects milk from its affiliated societies, *and also from other individual producers, pasteurizes it and sends it to Bombay through insulated railway vans.*" During the year 1949-'50, the union (KDCMPU) had collected 60,01,500 lbs. of milk, valued at Rs. 13,06,248 and realized Rs. 12,74,449 and thereby made a profit of Rs. 81,215 at the end of the year.

Notes

1. Calcutta had passed a first legislation in the year 1866, but it had had a limited reach and was applicable only within the municipalities in Calcutta. Mishra (2015: 111) recounts that "this was followed by the 1886 bill, which had a slightly wider jurisdiction; the 1886 bill, in turn became the template for a similar bill that was passed in Bombay in 1899". The Government of Bombay had passed another legislation (Bombay Act VI of 1913) which laid down processes for the sanitary control of milk supply of Bombay and required all the dairymen, dealers, importers or sellers of milk to hold licenses in Bombay.
2. O'Gorman, P. W. (1895) 'Dairy Produce in India', A paper read at the Indian Medical Congress, 1894; Caledonian Printing Works, Calcutta. pp. 03, 10 and 12.
3. Probably the impulse of establishing large-scale dairy farms by military authorities, starting from the first one at Allahabad in the year 1891, was also rooted in repeated articulations of such sentiments.
4. Preface to *Cows in India and Poultry: Their Care and Management* by E.B.J. Second edition of this book was brought out by the author in 1896, where she notes that the first edition was received very kindly by readers and several other authors had used extracts from it. It seems in second edition, few remarks on keeping poultry were also added.
5. Stow, A. M. (1910) *Cattle and Dairying in Punjab,*
6. Cited in Mishra, Saurabh (2015) *Beastly Encounters of the Raj: Livelihoods, Livestock and Veterinary Health in North India 1790–1920*, Manchester University Press, Manchester, p. 106.
7. While this tendency is witnessed in several books in this genre, for an example see, second chapter titled, 'Doodh naa Praachin Ullekh' (Milk in the Ancient Texts) in Shah, Narsimh M. (1940) *Dudh: Sarva SampurN Khorak* (in Gujarati), Gujarat Vernacular Society, Ahmedabad, pp. 06–10. This was adapted from Prof N. N. Godbole's famous 1936 book, 'Milk: The Perfect Food', which he had self-published and which appeared in translation in several Indian languages. When we look at acknowledgement, we realize that Prof Godbole had requested his colleague Dr B. K. Ghanekar to supply him references concerning milk from Ayurvedic texts and Pandit S. D. Satwalekar of the Oundh state had helped him with quotations from the Vedas.
8. See, Joshi, Lemuel Lucas (1915) 'Bacteriological Examination of the Bombay Milk Supply and the Question of Bacterial Standards for India,' *The Journal of Dairying and Dairy Farming in India*, Vol III, Part 1, pp. 05–36.
9. See Bose, Chunilal (1910) 'Necessary Measures for Prevention of Food Adulteration', Calcutta Medical Journal (July 1910).
10. In the year 1913, out of 593 samples examined, 40.1% were found to be adulterated, in 1914 out of 496 samples almost half were found to be adulterated and in 1915 once again 40.2% of the total 490 samples examined were found to be adulterated. In the year 1916, the prevalence of adulteration appeared to have come under check and just 26% out of 403 samples examined were found to be adulterated, but in the next year (1917), once again out of 436 samples examined the extent of those with adulteration turned out to be 40.6%. Thus on an average during this five year period 40% of samples examined in Calcutta Municipal Corporation laboratory were found to be adulterated.

11. Speaking at the same conference Dr. Srinivasa Rao reported that an analysis of 315 samples of milk in Bangalore showed that 78% of these samples were adulterated with water. Similarly, Harold Mann in his paper 'The Supply of Milk to Indian Cities' reported that he had found almost 80% of the milk supplied to Poona city adulterated with water. Mann qualified his remarks stating that "Unless 2 annas per pound were paid, it was impossible to get unadulterated milk" since "in exact proportion as less was paid, water was added". Proceedings of the Third All India Sanitary Conference held at Lucknow in January 1914, pp. 91–92 & 94. Similarly, a study by Rai Bahadur Dr. Chunilal Bose titled *The Milk Supply of Calcutta* (1918: 04) talked about an inquiry carried out in 1913 which revealed that 83.6% of the milk supplied to hospitals in Calcutta was adulterated. Bose, Chunilal (1918) *The Milk Supply of Calcutta*, Central Press, Calcutta.
12. This essay titled, 'Fixing of Standards of Purity of Milk and its Products' was published in December 1921 issue of Indian Journal of Medicine. See, Bose, J. P. (ed) (1924) *The Scientific and Other Papers of Rai Bahadur Chunilal Bose* (Volume II), Forward Press, Calcutta.
13. The next chapter presents the discussion on the introduction of organized dairying in India amongst colonial policy makers. The next chapter engages with the emergence of Military Dairy Farms, the question of dairy education in India and the emergence of dairy technology.
14. For a much detailed account of the spectacular effect of the adulteration cases regarding ghee, first in 1886 in Calcutta and then in 1917, see Mishra, Saurabh (2015) 'Food Adulteration, Public Health and Middle Class Anxieties' in *Beastly Encounters of the Raj: Livelihood, Livestock and Veterinary Health in North India 1790*–1920, Manchester University Press. Thanks to his discussion on how a vegetarian community responded to food adulteration incidents that appears to have ghastly consequences for their own sense of religious and caste purity, I tried to probe whether there were similar spectacular outrages in Gujarat.
15. Adhiya, Nandlal Chhaganlal (1925) *Vegetable Ghee ane PraNi Raksha* (In Gujarati), retrieved from archives.org.
16. The same author had also written a longer version of his arguments against *banaspati* in a book form on Gujarati under the title 'Vegetable Ghee no Itihaas' (The History of Vegetable Ghee), but he felt the need to bring out a smaller booklet in fifteen pages, priced at one and a half anna to lend his support to ongoing mobilizations.
17. Agriculture and Animal Husbandry in India for the year 1938–'39, p. 190.
18. Patel, Sujata (1990) 'Anand Model' in Doornboss, Martin and K. N. Nair (eds) *Resources, Institutions and Strategies: Dairy Aid and Operation Flood*, Sage Books, New Delhi.
19. When one looks at the number of plough and cart oxen available against the number of ploughs and carts, one finds that the number of oxen is actually in deficit. The ideal number would have been around 170,000, and thus the number that was being reported was around 40,000 less than the ideal agricultural stockholding. Sujata Patel makes a conjecture following from these figures that the Patidars in Kheda had started to share plough bullocks to prepare the farms ahead of monsoon rains and identifies that such a cooperative act of sharing the mobile forms of property owed to the great cohesiveness that is so characteristic of the Patidar community.
20. See, Hardiman, David (1981) *Peasant Nationalists of Kheda*, Oxford University Press, Delhi, p. 262.
21. See, Hardiman, David (1981) *Peasant Nationalists of Kheda*, Oxford University Press, Delhi, p. 262.
22. His grandson, Sarvadaman Patel continues practising farming since 1980 and has won several awards in recognition of his Organic Farming practices. *Bhaikaka Krushi* Kendra is a 40-acre integrated, holistic Organic & Biodynamic farm that incorporates all aspects of farming like vegetables, cereals, fodder, animals, agro-forestry, water harvesting, orchards, etc. The main focus of the farm is the dairy of nearly 100 animals & the production of a rich variety of fodder to feed these animals. The farm produces about 300–350 L of milk every day and has been catering to 85 customers and a 500-bed hospital in Vallabh Vidyanagar for last 25 years. The farm also produces a variety of fruits like chikoo, mango, custard apple, litchi, papaya, mosambi, and lime. Over 40 different vegetables are grown mostly in the winter months. Carefully designed trenches & ponds ensure that all the rainwater is harvested. A variety of

cereals & pulses like oats, bajra, barley, wheat are grown. For more details, see https://www.organichutbkk.com/.

23. Not only Charotar, but Bombay Presidency had also reported the largest acreage under bajra at 4.8 million acres. See, Russell, E. John (1937) Report on the Work of the Imperial Council of Agricultural Research in Applying Science to Crop Production in India, Government of India Press, Simla p. 30. Hardiman (1981: 261–262) points out that the staple diet for peasants was chapatti made from bajri and juwar, two types of millet. He underlines the important of Bajri by underlining that in the year 1910–'11, 26% of the district was under bajri, 10% under juwar, Kodra (16%) and Ragi (10%) were now being seen as "inferior cereals" that was grown and eaten by poor peasants. Lentils (12%) cooked in the form of Daal provided the peasant's protein. Rice (12%) and wheat (2%) were considered luxury cereals, grown usually for sell and eaten only by the rich or given to the honoured guests. The other important crops in Kheda district were tobacco (4%), cotton (3%) and mhowra, which came from the trees. However, Hardiman also underlines that during the early twentieth century, there was a great expansion in tobacco and cotton cultivation. While in the year 1897–'98, only 4% of the cultivated area of the two Charotar talukas of Anand and Borsad was under tobacco, by the year 1921–'22, the figure had risen to 13%. Similarly, after the year 1915, when it was discovered that good cotton could be grown in the Mal tract, during the next decade, cotton became the major crop of that region of Kheda district.
24. Bhailal Patel (born 1888—died 1968) recollects this part in the first chapter of his memoirs, Bhaikaka naa SansmaraNo, that got published after his death in 1970. See, Patel, Bhailal (1996, second edition) Bhaikaka naa SansmaraNo, Sastu Sahitya Vardhak Karyalay, Ahmedabad. p. 12.
25. Bates, Crispin (1981) 'The Nature of Social Change in Rural Gujarat: the Kheda District, 1818–1918,' Modern Asian Studies, Vol XV, No 4, pp. 771–821.
26. Joslen, F. (1905) Cattle of the Bombay Presidency, Bulletin No 26, Department of Land Revenue and Agriculture, Bombay.
27. Editorial of The Journal of Dairying and Dairy Farming in India, Vol IV, Part 1, October 1916, p. 04.
28. Hardiman, David (1981) Peasant Nationalists of Kheda, Oxford University Press, Delhi, pp. 123–125.
29. For details of his arguments against the decision of closing down see, Royal Commission on Agriculture, Vol I, Part II, pp. 16–17.
30. Review of Agricultural Operations in India for the years 1929–30 & 1930–'31, pp. 205 & 206.
31. Agriculture and Animal husbandry in India 1934–'35 and 1935–'36, ICAR, 1936, pp. 03 and 21.
32. Agriculture and Animal Husbandry in India for the years 1933–'34 and 1934–'35, p. 03. Around this period of mid-1930s, another proposal was also put forward to create a separate office for dairying under the Government of India.
33. For Arthur Olver's note on Dr. N. C Wright's report see, Annexure 3, pp. 330–335.
34. See, Warner, James (1951) *Dairying in India*, Macmillan & Co Ltd, Calcutta, p. 318.
35. For the myth of origin of Calcutta Milk Supply Union, see, Donovan's paper titled, 'Cooperation and Milk Supply of Calcutta', which was published by the Bengal Cooperative Organisation Society. As per this origin myth, recounted in a 1924 note penned by Babu Akshoy Kumar Bose, Inspector of Cooperative Societies, Calcutta, "In the course of his tour, Mr Donovan came across a Mahommedan cultivator who had boycotted the local *goala* for the latter's refusing to come to one of their marriage feasts and the Mahommedans of the village wished to supplant the *goala* by one of themselves for taking the village milk to Calcutta" and had found the task a bit daunting. Readers are told that "Mr. Donovan at once seized the opportunity and accompanied the cultivator to his village and arranged for the sale of milk in Calcutta, under the aegis of the Cooperative Department". Thus, the first milk producers' cooperative society got started by sending a meagre 15–20 seers of milk a day to Calcutta in the year 1915. This note by Babu Akshoy Kumar Bose was reproduced in an article by Rai Bahadur Chunilal Bose titled, 'The Milk Supply of Calcutta: Its Hygienic Commercial

and Social Aspects', published in Bose, J. P (ed) (1925) *the Scientific and Other Papers of Rai Bahadur Chunilal Bose*, Forward Press, Calcutta. p. 144. While the myth of origin may allude to the cooperative milk supply union to the year 1915, the union got formally registered later than that date. As per another narrative, the first milk cooperative society was formally started in April 1917 at Baraset and the success of this society was followed by establishment of five others in the next year, and their number increasing to nineteen in the year 1919, when Calcutta Milk Supply Union got formally registered. See, Niyogi, J. P. (1940) *The Cooperative Movement in Bengal*, Macmilan. pp. 223–224.

36. This paragraph is based on Royal Commission on Agriculture in India, Abridged Report (1928: 228).
37. All the figures as reported in Review of Agricultural Operations in India for the years 1926–'27, p. 88.
38. There has been a reference of Calcutta Municipal Corporation subsidizing the cost towards establishment of pasteurization and bottling plant of the union in Royal Commission on Agriculture in India, Abridged report (1928: 228). Another contemporary source, Coatman (1928: 340–341) states that, "The Municipal Corporation of Calcutta granted to the union a non-recurring subsidy of Rs. 5000 and a loan of Rs. 50,000 free of interest, repayable in six years, on condition that the union increased its yield to 500 mds a day within three years. With this assistance, the union was able to erect a dairy with up-to-date plant for the purpose of pasteurizing and treating milk". See, Coatman, J. (1928) *India in 1926–'27: A Statement prepared for Presentation to Parliament in accordance with the requirements of the 26th Section of Government of India Act*, Government of India, Central Publication Branch, Calcutta.
39. The cost is in Rs-Ana-Pie format.
40. Review of Agricultural Operations in India for the years 1929–30 & 1930–'31, p. 206.
41. Review of Agricultural Operations in India for the years 1926–'27, p. 88.
42. One Madras Measure equals four pounds of milk. An article that reviewed the work done by Madras Cooperative Milk Supply Union in the year 1948 stated that "the Union handles about 5000 Madras measures at day at present". See, Naidu, K. Muthukrishna (1948) 'The Madras Cooperative Milk Supply Union', *The India Cooperative Review,* Vol XIV, No 03, July–September 1948, pp. 175–180.
43. See, Naidu, K. Muthukrishna (1948) 'The Madras Cooperative Milk Supply Union', *The India Cooperative Review,* Vol XIV, No 03, July–September 1948, p. 177.
44. See, Warner, James (1951) *Dairying in India*, Macmillan & Co Ltd, Calcutta, p. 319.
45. See, *Annual Report of Cooperative Societies, 1928–'29*, p. 29–30.
46. See, Shah, K. T. (ed) (1948) *Report of Sub-committee on Animal Husbandry and Dairying*, National Planning Committee Series, Vora and Co Publishers Ltd. Bombay, pp. 73–74.
47. See, Mukherjee, B. B. (1937) *Agricultural Marketing in India*, Thacker Spink & Co Ltd, Calcutta.
48. See, Hough, Eleanor M (1932) *The Cooperative Movement in India*, P. S. King & Son, Westminster, London.
49. An article that reviewed the work done by Coimbatore Cooperative Milk Supply Union in 1948 reported that the Union now had 26 affiliated cooperative societies, twelve of which were organized at considerable distance from Coimbatore city–around 30–50 miles–and were largely the outcome of the need to pool milk to meet military demands during the war years. These cooperative milk producers' societies around urban settlements such as Tiruppur, Dharapuram, Udamalpet, Palladam, Avanashi and Pollachi operated as independent supply societies catering to population in these urban centres, (with exception during the war years) but they still sought to affiliate themselves with the Coimbatore union in order to receive the guidance. This 1948 review article stated that only "six out of 26 societies are said to be dormant". See, Ramakrishnan, K. C. and V. S. Narasimhan (1948) 'The Coimbatore Milk Supply Union', *The India Cooperative Review,* Vol XIV, No 03, July–September 1948, pp. 181–185.
50. An article that reviewed the work done by Lucknow Cooperative Milk Supply Union stated that "During the decade beginning from 1938–'39, the Union made profit in the first five years, but suffered losses in the remaining five, with a net loss of Rs. 499 at the end of

the decade" and considered financial position of the Union in 1948 as far from satisfactory. Talking about the Government support, it reported that "Practically no substantial pecuniary or other help was received during the regime of Advisory Government". Despite this, one doesn't come across a very despondent tone in the article which was written in the year after India winning independent. Great hopes were pinned upon the newly established National Government which was found to be "most anxious to increase the production of milk all round by giving substantial help". Replication of the Calcutta model of 'organising city milk supply on cooperative lines" was a task that was taken up at once by the Government action of setting up a separate Dairy Section in the cooperative department with the requisite gazetted and non-gazetted staff and Cooperative Milk Supply Union had been established at Allahabad, Banaras and Kanpur and few more were in the process of being established at Faizabad, Jhansi, Nainital and Meerut. Thus in 1948, Lucknow Cooperative Milk Union was promised "free supply of the pasteurizing plant and other equipments of the value of Rs 1,53,000 and funds amounting to Rs 64,000 for the construction of milking sheds etc." See, Lal, Gopal (1948) 'The Lucknow Cooperative Milk Supply Union', *The India Cooperative Review*, Vol XIV, No 03, July–September 1948, pp. 168–174.

51. However, it shall be kept in mind that the examples that are cited by Wright are of those places that he had been able to visit personally. So, the possibility of other such similar success story village cooperatives and co-operative dairies beyond those names could have existed and the list shall not be taken as being exhaustive.
52. See, Naidu, K. Muthukrishna (1948) 'The Madras Cooperative Milk Supply Union', *The India Cooperative Review*, Vol XIV, No 03, July–September 1948, p. 177.
53. *Annual Report on Working of Co-operative Societies in Bengal for the year 1944*, Government Printing Press, Alipore, Bengal p. 06.
54. *Annual Report on Working of Co-operative Societies in Bengal for the year 1945*, Government Printing Press, Alipore, Bengal p. 68.
55. Biswas, James Nirendra Kumar (1948) 'Calcutta Cooperative Milk Societies Union', *The Indian Cooperation Review*, Vol XIV, No 03 (July–September), p. 165.
56. Amongst several conjectures that one can hazard is the bloodied partition of the country in 1947, with creation of East and West Pakistan. As the annual report for 1945 shows there was a Chittagong Central Cooperative Milk Supply Union which after August 1947 had become part of East Pakistan and wouldn't have kept marketing linkages. Also, the large-scale movement of people with their livestock from both sides of border would probably have meant that village level cooperatives suddenly lost many members who were pouring milk into Union's milk collection centres. While the references to Calcutta Milk Union and other four milk unions become rare after late 1940s, on 9th June 1958, in the West Bengal state assembly, it was reported that "We have 174 primary dairy cooperatives in different parts. The total number of members is 12138; they sold Rs. 2.94 lakhs worth of milk and Rs. 3.8 lakhs worth of milk products. There is one federated Calcutta Milk Union which sold last year Rs. 6.67 lakhs worth of goods".
57. See, The Report of Cooperative Planning Committee, 1946, p. 42.
58. Heredia claims that by the year 1910–'11, Stafford was sending about 45,500 kg of cream to Bombay every day.
59. Heredia claims that by the year 1916–'17, Indian Dairy Supply Co. was transporting approximately 11,365 kg of butter every day to Bombay, and only a little less than that amount of pasteurized milk, besides Paneer.
60. Cited in Chatterjee, Nilanand (1924) *Condition of Cattle in India*, All India Cow Conference Association, Calcutta.
61. See, Royal Commission on Agriculture, Volume, Evidence in Bombay Presidency, p. 228.
62. Annual Report of the Department of Agriculture, Bombay Province for the year 1936–'37, pp. 30–31. Out of the total 2072 acres area that government of Bombay had earmarked for Palghar Dairy District Scheme, the two land plots were leased out to Mr. M. M. Killedar (500 acres) and Mr. Haji Moosa Umarji Bharucha (537 acres). More applications from milch stable owners were pouring in during the year 1936–'37.

63. Annual Report of the Department of Agriculture, Bombay Province, 1937–'38, p. 31.
64. Annual Report of the Department of Agriculture, Bombay Province, 1944–45, p. 27.

Chapter 6
Discourse on Dairy Development, Scientific Instruction and Emergence of Technocracy

Norman C. Wright (1937: 08) considers that efforts to organize dairying during colonial period had twofold origins, "First, the introduction of cream separators in 1889, and second, the establishment of large-scale dairy farms by the military authorities; the first of which was started at Allahabad in 1891".[1] Of these two developments, the latter was to prove the most important since it led directly to the creation in 1920 of the post of Imperial Dairy Expert. In addition to these two early efforts, in the year 1913, a professional organization, namely the Dairy Education Association of India started publishing its journal titled, 'The Journal of Dairying and Dairy Farming in India'. Two subsequent meetings of the Board of Agriculture in India—the eighth meeting held at Coimbatore in December 1913 and the ninth meeting held at Pusa in February 1916—had also discussed the subject of organizing dairy development in India.

Towards the last decade of the nineteenth century, private entrepreneurs had started to explore whether commercial scale dairy farming could do well in India. A Swiss gentleman Mr. Edward Kaventer, who had accompanied Mr. Howman—a well-known dairy farmer from Warwickshire—who visited India in 1889 on behalf of the Dairy Supply Company Limited for the purpose of introducing mechanical cream separators, was retained in India by the Bombay Government. Under his guidance, the Agriculture Department started a working dairy plant fitted with cream separators, churns, refrigerators, etc. Voelcker (1893: 208) mentions that Mr. Kaventer also worked with the Government of North West Provinces for some time at Cawnpore. Realizing the fortunes in dairy trade, Kaventer started his own private firm near village Cheerat in Aligarh district in 1897. His firm became so successful in the dairy business that soon he was able to expand his dairying business by commanding a considerable share of Calcutta and Shimla milk market. Similar to Edward Kaventer, another Swede, A. C. Stafford started the first modern dairy plant under the name British Express Dairy in Kheda district around the year 1900.[2]

Colonel A. S. Marriott's submission before the Royal Commission on Agriculture suggests that in 1889 Military Dairy Farms were opened for the purpose of providing

H. Upadhyaya, *Critical Insights on Colonial Modes of Seeing Cattle in India (1850–1980)*, Asia in Transition 27, https://doi.org/10.1007/978-981-97-1560-2_6

Table 6.1 Quantities of dairy products sold at Military Dairy Farms

Year	Milk		Butter	
	Quantities	Average sale rate per lb. (Rs-A-P)	Quantities	Average sale rate per lb. (Rs-A-P)
1921–'22	5910 tonnes	0–3–10.4	509 tonnes	1–11–8.5
1922–'23	4519 tonnes	0–3–6.5	450 tonnes	1–9–10.5
1923–'24	3839 tonnes	0–3–4.2	343 tonnes	1–9–8.4
1924–'25	3572 tonnes	0–2–11.5	341 tonnes	1–8–8.4

Source Marriott, A. S. (1928) Written Submission by Col A. S. Marriott before the *Royal Commission on Agriculture*

a safe and reliable supply of pure milk, butter and cream to military hospitals and troops in cantonments. Each Military Dairy Farm was accompanied by Grass and Fodder farm, but in the year 1910, Government of India not satisfied with the general economic workings of dairy and grass farms, appointed a committee composed of Mr. S. H. Fremantle, who was ICS and Lieutenant Colonel T. J. Henderson of Military Accounts Department. The committee recommended separation of the dairy farms from grass farms. In 1906, seventeen years after the opening of Military Dairy Farms, the service of dairy expert were obtained from England by hiring Mr. W. Smith. The importation of exotic bulls had commenced in the year 1907, and initially, Military Dairy Farms regarded this measure as purely experimental. At first, it was thought that the bulls imported from Australia would be more likely to withstand Indian tropical climate than the bulls imported from the UK. However, the idea turned out to be fallacious and the stock was distinctly inferior. It was only around the 1920s that Military Dairy Farms had started importing British Friesians to continue with crossbreeding on experimental basis. By the year 1925, the number of adult stock with Military Dairy Farms was 3345 while young stock numbered 872 and there were 401 calves.

Marriott further clarified that at the end of 1924–'25, there were 1139 crossbred cows and 1214 country-bred cows. Marriott gave the figures on the quantities of dairy products sold during the year 1924–'25, compared with three previous years and average sale rates thus[3] (Table 6.1).

Wright (1937: 08) reproduced the figures of Military Dairy Farms milk consumed by British and Indian troops and concluded that there existed a wide gap between the dietary habits of Europeans and Indians. Wright had also cautioned against using the policy adopted at the Military Dairy Farms as a model for the development of dairy industry in India as a whole (Table 6.2).

S. S. Milligan, Imperial Agriculturist, had stated in a note written for the ninth meeting of Board of Agriculture (1916: 107) that the Military Dairy Farms had butter factories at Peshawar, Ambala and Ahmedabad, with Ahmedabad factory handling 1 tonne butter per day in busy season.[4] He also listed the well-equipped and carefully managed dairy farms at the following stations: Peshawar-Cherat, Rawalpindi-Muree,

Table 6.2 Milk and milk products consumed by British and Indian troops

	Milk (lb.)	Butter (lb.)	Cream (lb.)
British troops	13,095,577	1,041,599	194,157
Indian troops	3,965,817	688	36

Source Wright, Norman C. (1937) *Report on the Development of Dairying and Animal Husbandry in India*

Lahore, Dalhousie, Ferozepore, Jullundur, Sialkot, Karnal, Agra, Lucknow, Allahabad, Cawnpore, Jubbulpore, Mhow, Quetta-Ruk, Poona, Ahmedabad, Belgaum, Bangalore and Aden.[5]

Parallel to Military Dairy Farms, there was thinking amongst the Department of Agriculture officers of the Government of India that much was needed to be done on the subject of 'cattle breeding and dairying'. Thus, at the meeting of the Board of Agriculture at Coimbatore in the year 1913, the subject came under extensive discussion. In a 'Note on Dairying in India' written for this meeting, S. S. Milligan and C. M. Hutchinson (1913: 174) wrote that "although there is a little authenticated information on actual milk yields of this country, all authorities are agreed that the average of even the best breeds of cows does not exceed 2400 lbs. per lactation period and that of buffaloes 3600 lbs". This state of affairs was accompanied by the prices of dairy produce that were as high in India as they were within a few miles of London, in spite of cheap fodder and food grains, cheap labour and the absence of sanitary regulations. Thus, they arrived at a conclusion that "there is little hope of improvement until something has been done to improve milking capacity of the cattle". They also subscribed to the then prevalent opinion amongst many that "the town dairies are producing a serious drain on the best milking cattle in the country"; hence "the question of preservation by government agency of the best milking cows in the country is therefore an urgent one". They also felt that the "steps should be taken at an early date to improve the milking capacity of at least the cows, and if possible, the buffaloes, by selection or crossing".

They had also voiced a viewpoint, which considered breeding for milking qualities by selection in the absence of pedigree and milk records a lengthy and time-consuming process and gave an example of Pusa herd of the Sahiwal cows that had been in existence for over eight years and wherein there were only three really good milkers in a herd of 200 cows. However, they had also noted that "the Department of Military Dairies had ransacked the Punjab for good Sahiwal cows". This somewhat contradictory statement crept in, as they failed to explain why Military Dairies had so overwhelmingly taken to procure Sahiwal cows from Punjab. Other references from the period 1891–1913, such as Joslen (1905) and Hewlett (1909) also state that Military Dairy Farms were looking for high milk-yielding breeds like Kaknkrej, Gir, Sindhi, etc. amongst Indian cows that went into making Military Dairy herds. Thus, after taking just one example of Pusa herd, Milligan and Hutchinson (1913: 174) concluded:

> It is therefore apparent that improvement through selection will be a long and difficult task *and it seems certainly worthwhile to carry out crossing experiments with imported sires*

> *from good dairy breeds*. The Military Dairies have already shown the way through the introduction of Ayrshire blood into their milking herds and have so far as they have gone, been quite successful. A scientific study of the inheritance of the important characters of dairy cattle by crossing might lead to valuable results. *This could be undertaken at Pusa with a certain amount of help by the Provincial Agricultural Departments in the way of using half-bred bulls on those farms where milking herds are kept.*

In addition to this note by Milligan and Hutchinson, a committee chaired by Mr. Blackwood studied 'The Dairying Industry in India and How Should the various Scientific and Commercial Problems Involved, be Attacked' and submitted its report.[6]

At the Coimbatore meeting (1913: 50), following resolutions were passed:

> Resolution No XV: That the board endorse the recommendation of the committee appointed to consider the question of Dairy Industry in India that Dairy as a branch of Agriculture Science should in future occupy a prominent place in the programmes of the Agriculture Departments.

This was followed by another resolution numbered XVI in the proceedings that stated:

> That the Board accepts the general principles enunciated in the report of the committee to consider the question of Dairy Industry in India and desire to emphasise the importance of a satisfactory solution of the problem of milk supply of towns, not only in itself but from the point of view of stopping the drain of good milch cows and buffaloes to towns where they are not utilized for breeding purposes to the best advantages. The Board considers that some legislation is necessary.

This early discussion on the State efforts for Dairy Development and Milk Supply to Cities was followed by further discussion between officers of Agriculture Department and Military Dairy Farms which led to drawing up of a partial scheme for the improvement of cattle breeding and dairying. In a memorandum placed for discussion by the ninth meeting of Board of Agriculture held at Pusa from 7 to 12 February 1916, charting out a general policy for the improvement of cattle breeding and dairying, B. Coventry, Agricultural Adviser to the Government of India (1916: 102) wrote:

> (a) Measures for the protection and amelioration of the existing cattle breeding industry, the preservation and multiplication of the existing fine types of Indian cattle, the organization and regulation of breeding tracts.
> (b) The preservation of grazing areas, the improvement of waste areas, and the general question of fodder supply.
> (c) The improvement of Indian cattle by breeding and selection on a large scale for the production of pedigree stock *with the object principally of increasing the yield of milk and the setting up of a good milch cow*. (emphasis supplied).
> (d) Crossbreeding.
> (e) The erection of model dairies throughout the country to aid (c) and (d), for obtaining milk records and for the spread of modern methods of dairying.
> (f) The study and improvement of existing indigenous method of treating milk and its products.
> (g) The erection of dairy schools attached to dairies for training and education.

(h) The fostering of dairying as a business on commercial and cooperative lines for the supply of milk and milk products to the public.

(h) The application of sanitary control over the supply of milk and its products.

Coventry acknowledged immediately after outlining this general policy that "it would be difficult to take up at once all the problems enumerated" and considered it "satisfactory if a beginning were made in such items which might be suitably taken up" and called for the cooperation from the Provincial Departments of Agriculture, Cooperation, Veterinary and Sanitation. Coventry's memorandum had suggested that Imperial Expert Breeding and Dairying Officer to the Imperial Department of Agriculture be appointed with headquarters at Kasauli. In addition to this officer, Coventry (1916: 103) suggested that a Chemist shall be appointed to work along with him to carry out "work on the value of food stuffs and the digestive capacity of Indian animals". He also pointed out that measures will have to be taken for the prevention of Rinderpest and other diseases amongst imported and cross-bred cattle owing to their high susceptibility to disease and suggested cooperation from Imperial Veterinary Research Institute, Muktesar staff for the same. As per this scheme, the imperial expert was supposed to carry out the cattle breeding work in coordination with Military Dairy Farms such as Lahore-Ferozepur, Lucknow, Poona, Belgaum, Ambala, Jubbulpore and Quetta-Ruk; crossbreeding experiments were proposed only at Amballa and Bangalore; academic courses on dairying were supposed to begin with establishment of Dairy Schools at Bangalore, Poona and Lucknow, and experimental work by physiological chemist to determine food value and digestive capacity of Indian animals was to be attempted at Pusa and at different Military Dairy Farms.

This memorandum was accompanied with 'Notes on Recommendations of Board of Agriculture in 1913, regarding the Dairy Industry in India' penned by S. Miligan, Imperial Agriculturist (1916: 105–106) who reiterated the Board position stating that "the Board laid great stress on the point of view that the production of a dual purpose animal should be the aim of Government Cattle Breeding Operations in India". He acknowledged that in certain quarters, there existed a view that a dual purpose animal for milk and draught is unobtainable, but pointed out that such was not the case as proved out both by experience and by crossing experiments. Miligan talked about the crossbreeding experiments by Military Dairy Farms in mating Montegomary cows with Ayrshire bulls and producing females with excellent milking qualities and males of exceptional draught power. He had also pointed out that while cattle in India on the whole developed such distinctive characteristics that they got classified into milch and draught breeds, he had come across individual animals carrying both qualities. Milligan considered dairy farms very essential for breeding dairy cattle, since he stressed the role played by milk recording in the process of selection and felt that this could be done economically only at dairy farms. He had also stressed upon experimental crossings with a view to obtaining milk pedigree. He also spoke about three breeds of cows and a buffalo breed that had animals giving high yields, namely the Hansi-Hissar, the Sindhi and the Sahiwal (Montegomary) cattle and the Delhi strain of buffaloes.

Milligan had pointed out that in order to carry out such a scheme as pointed out by Coventry and him, the cooperation of the Military Dairy Farms Department would prove invaluable. He reminded that at the Coimbatore Conference in 1913, the Director of Grass and Dairy Farms had indicated the willingness to assist in the direction of instruction and experiments. Milligan then went on to outline the scheme of distribution suggested by Mr W. Smith, Assistant Director, Dairy Farms, Southern Circle. Smith (1916: 106) in his scheme of distribution suggested rearing bulls of Sahiwal breed for distribution at Lahore and Ferozepur Military Dairy Farms; suggested establishing a herd of 100 best Hansi-Hissar cattle from the Military Dairy Farms at Agra, Cawnpore, Allahabad and Lucknow to be established at Lucknow; another herd of pure-bred Sind cows at Government Dairy Farm, Ruk junction in Queta; another herd of pure-bred Sind cows at Belgaum Military Dairy by drawing upon animals from Poona and Belgaum together. While Coventry's scheme had suggested crossbreeding of Indian stock with imported bulls of milking strains only at Ambala and Bangalore, Milligan's note suggested such experiments at four Military Dairy Farms at Ambala, Jubbulpur, Poona and Bangalore. Milligan had desired that the question of grading up of remaining breeds of cattle in the country could be taken up by Provincial Agricultural Departments and expected Pusa to take lead in this task. Milligan had also suggested milk recording societies throughout the country and to establish six properly equipped dairy schools in the Punjab, the United Province, Bengal, the Central Provinces, Bombay and Madras. By the year 1917, the consistent work of milk recording societies in Sweden had shown the following calculated annual yields per cow: 1870: 2600 lbs; 1880: 3700 lbs; 1903: 4200 lbs; 1913: 4730 lbs.[7]

Very early on S. S. Milligan (1916: 107) had voiced his opinion that "The distribution of pedigree bulls, the establishment of milk records, the spread of knowledge amongst the people, the organization of dairy concerns and the marketing of dairy produce can be best done through cooperative societies".

At the ninth meeting of Board of Agriculture that took place at Pusa (1916: 26), the subject of 'cattle breeding and dairying in India' was discussed by a committee chaired by Mr S. S. Milligan. The committee's terms of references read as follows:

> (i) To examine the scheme for cattle breeding and dairying as set forth in Memorandum by the Agricultural Adviser to the Government of India, dated 19 July 1915, and to make recommendation.
>
> (ii) To examine any other schemes, proposals or aspects before the Board and to make recommendations.

At the ninth meeting of the Board of Agriculture held at Pusa in February 1916, Mr. Miligan (1916: 26–29) read out the report of the committee and then moved the following resolution:

> That the Board is of opinion that in order to make satisfactory progress in the development of good breeds of milk cattle in India and in Dairying, an Imperial Expert Breeding and Dairying Officer should be appointed with the title of Imperial Dairy Expert whose practical duties would be, (a) the control of the cattle breeding farms and dairy operations contemplated in the scheme; (b) the supervision of dairy instruction; (c) the study and improvement of

> existing dairy methods in the country and the establishment of the industry on commercial basis; (d) generally to advise and assist Local Governments, Provincial Officers, Military Dairy Farms. They consider that the arrangements proposed and the estimate prepared by the Committee are reasonable and that the holder, once appointed, should not be liable to transfer.

However, several changes were proposed when this resolution was moved. When Mr. Townsend enquired whether the expert would have to deal with plough cattle, Colonel Pease (1916: 30) suggested that "he did not think it possible to get any of the very fine breeds of draught cattle in India to become also milch breeds without deterioration into draught qualities" and stressed that he was voicing this opinion as the result of 31 years of experience. He also felt that "the proposed officer should not be an expert in breeding draught stock and he thought the scheme should be limited to milch cattle". Dr. Mann also joined his voice along with Colonel Pease in pointing out that "resolution referred to milch cattle and breeding draught cattle is for provinces to decide and not for the imperial expert". These words by Dr. Mann suggest that in the thinking amongst colonial experts, there existed a crucial division of labour: draught bullock requirement of peasants was to be looked at by provincial experts, while organizing dairying in India on modern lines was a project worth the attention by imperial experts!

At the ninth meeting of Board of Agriculture, Major Walker explained that the scheme outlined by Milligan had stressed upon the help, Department of Agriculture was to avail from Military Dairy Farms and moved the following resolution:

> That the Board is of opinion that the offer by the Military authorities of the herds of various breeds of pure bred Indian cows and buffaloes, as well as of the facilities for conducting further breeding work on the Military Dairy Farms is of extreme value, and should be gladly accepted. They recommend that advantage should be taken of it as soon as the Imperial Dairy Expert referred to in the foregoing resolution is appointed.

6.1 Appointment of Imperial Dairy Expert (1920)

By the year 1916, there were committees in several cities like Calcutta, Bombay and other large towns in India that had discussed the question of organized dairying, Military committees had on several occasions had considered the development of milk supply to cantonments and the All India Sanitary Conference on two occasions had this subject on its agenda. In response to recommendations made by the Board of Agriculture at their meetings at Coimbatore in 1913 and Pusa in 1916, the Government of India in May 1920 appointed an Imperial Dairy Expert and in July 1921, the appointment of a physiological chemist was made to carry out research work, in connection with problems of animal nutrition. The Board of Agriculture at the above-mentioned meetings had also recommended the establishment of dairy schools and of breeding farms on moderate scale, but government did not provide either money or facilities of any sort to Imperial Dairy Expert to carry out any educational or research work until July 1923.

It also becomes very clear from the written submission by Imperial Dairy Expert, Mr W. Smith to Royal Commission on Agriculture that while three Military Dairy Farms were handed over to him for research and education, these were needed to be run commercially well, and there were no additional monetary grants for research and education that came from Government of India. It also becomes very obvious from the oral evidence he submitted to Royal Commission on Agriculture (1927: 25) that he was considerably disturbed by the fact that he practically lost three years after his appointment as Imperial Dairy Expert[8]:

> What happened during the gap of three years between your appointment as Imperial Dairy Expert and the taking over of your farms? – *Nothing happened. We had no funds and no facilities to do anything. I used to write to Mr Milligan almost every day for three years and then we got these three Military dairy farms.*
>
> How did you keep yourself employed? – By giving advice and drawing up plans for people, things like that; we did what we could; I think the time was wasted.
>
> So that you really lost three years? – Yes. (emphasis added)

As soon as Imperial Dairy Expert, W. Smith was handed over the three Military Dairy Farms; he had to work out a cattle breeding policy to be adopted at these farms and guiding him in this task was his prior work with the Department of Military Dairy Farms. Thus, the first Cattle Conference (1924: 06) discussed the cattle breeding policy to be followed at these three farms at Bangalore, Wellingdon and Karnal as well as for a herd maintained at PUSA, Delhi.

In the year 1923, when the first Manual of Dairy Farming for Indian Students got published, its author B. K. Ghare wrote in preface, "Dairy literature published in our country is very poor and this fact alone is quite sufficient to show how far our country is lagging behind other nations in the science and art of dairying".[9] Within five years, this manual had its second revised edition published with an introduction by William Smith, Imperial Dairy Expert and Secretary to Central Bureau of Animal Husbandry. Smith had called the second edition of this manual by Ghare, "the only modern publication which attempts to deal with the dairying problem from an all-India point of view" and had hoped that "it will meet an ever-increasing demand as a textbook and reference work regarding Indian Dairying".[10] By 1923, at the second conference of veterinary officers, Mr. Edwards mentioned that there existed suitable facilities for imparting some practical training to students on cattle keeping. Edwards listed the herd of dairy cattle at Pusa maintained under the supervision of the Imperial Agriculturist and the well-known breeding herd of draught cattle at Hissar.[11]

6.2 Milk Recording Societies

Milk recording and pedigree registration were believed to be the most important aspect in efforts in the direction of organized dairying. There was a growing realization during mid-1930s that the countries that had taken to them with earnestness represented the cattle breeding industry in most advanced status as detailed in data that started to get published by International Institute of Agriculture. At the Animal Husbandry Research Workers' Conference in New Delhi held in February 1936, E. J. Bruen had submitted a note for the discussion titled, 'A Note on the Establishment and Maintenance of Pedigree Herd Books in Bombay Presidency'. It seems by the year 1939, official herd books had been opened for some of the indigenous breeds maintained at government institutions and private owners were being encouraged to maintain records of yield, breed, length of lactation, etc. of their animals.[12]

Milk recording activities had started in six villages in Sanand taluka, Ahmedabad district during the year 1938–'39. In the next year, i.e. 1939–'40, the Government of Bombay "accepted the offer of the ICAR to finance half the cost (i.e. Rs. 960 at the then prevailing prices) of the village milk recording work in the Kankrej tract of the province for a period of two years". Subsequently, "a milk recorder was appointed on 17 May 1939 and was given a practical training at the Chharodi Cattle Breeding Farm for a period of fortnight in milk recording, testing, etc."[13] Thereafter, this trained worker was put in charge of Milk Recording Societies in nine villages in Sanand taluka, "where Premium Bull work had been in progress for 15 years" and the cooperation from Kankrej cattle breeders was sought for this scheme.

The first progress report of the Milk Recording Scheme of Bombay province was discussed at the meeting of the Milk and Milk Products Committee of the Advisory Board, ICAR held in New Delhi on 15 November 1940. The committee expressed the view that the scheme had proved to be of great value in connection with the Provincial Pedigree Registration Scheme already in progress in the same area.

During the year 1941–'42, this scheme continued to be in operation "in nine villages in Sanand taluka, Ahmedabad district, and the milk yields of 143 cows were recorded and quality determined in a series of fortnightly tests". The scheme had also been granted an extension for the further period of one and half years (i.e. till 31 March 1943). In the year 1944–'45, the milk recording was being carried out in 15 villages of Sanand and Viramgam talukas of Ahmedabad district, and the milk yields of 135 cows were being recorded.

Around 1938–'39, it was "believed that if the propaganda work (on profits to be made by breeding milch cattle) continued uninterrupted, the Annamalai areas will become a good breeding place for milch cattle". The extension works here comprised of demonstration on milk separation, milk testing and butter making under modern conditions.

The issue of development of pedigree registration and milk recording was on the agenda of the third meeting of the animal husbandry wing of Board of Agriculture and Animal Husbandry in India that took place in February 1939. The board was of the view that "pedigree registration and milk recording were of great importance in the improvement of milch breeds of cattle". Hence, the participants "stressed the essential point that milk recording must be conducted with great accuracy and uniformity of the method" as it greatly influenced the position of pedigree registration". The board "recommended that all methods of encouraging milk recording, such as establishment of milk collecting centres, etc. should be assisted in every way possible".[14] However, some indigenous breeds of the Montegomery and Kathiawar districts, Nili, Surati and Ongole buffaloes and the goats of Punjab, North West Frontier Provinces and

Bihar had not yet found their place in the official registers. In 1939, another bulletin published by Imperial Council of Agricultural Research presented the only reliable data on the milk yields of cattle managed under village conditions reported from seven principal dairy tracts of India.[15] Various steps in the direction of opening herd books were taken between and pan-Indian proper herd books actually started in the year 1941 only.[16]

6.3 Formation of Breed Societies

A breed society having as its main object the promotion of the interest of the famous Kankrej cattle breed of North Gujarat was established during the year 1941–'42, with eighty members coming together. A total of 628 Kankrej cattle had been registered, and "a scheme to combine herd book registration with milk recording under the auspices of the society" was submitted to ICAR during the same year.

6.4 Emergence of Technocracy

The ninth meeting of the Board of Agriculture held at Pusa on 7 February 1916 and subsequent days, had discussed the question, 'Which one is the best agency for controlling cattle breeding' and a committee had been appointed whose terms of reference were:

> To consider and say, which of the two Departments – the Agricultural and Veterinary – is, in the opinion of the Board, the better agency for controlling cattle breeding.

Keatinge (1916: 18) read out the report submitted by the committee whose main points were following:

> Looking to the existing cattle breeding organization in India we consider that it would be very unfortunate if either the Agricultural or Veterinary were to be entirely disconnected with cattle breeding or with the wider questions comprised in the term 'Animal Husbandry'. In connection with animal husbandry the problems appear to differ considerably in different Provinces, and the organization has developed on different lines. Where the problem is merely the improvement of particular breeds, the man with most experience of those breeds is the man best suited to control the work. Where the work or projected work is also connected with problems of fodder supply and other economic factors it is essential that the man in control should be well acquainted with the people and with the local conditions. Where the work is connected with the cattle-keeping as part of mixed farming it assumes a more definitely agricultural tinge; and this is still more the case in the matter of dairying. In controlling large cattle farms like Hissar, the question of contagious disease becomes of primary importance and veterinary knowledge is essential. The organization of the Agricultural and Veterinary Departments varies much in the different provinces. In some provinces there is a strong Veterinary staff well acquainted with the Province, while in others this is not so. *We therefore consider that it would be best for the various Provinces to arrange for the control of animal husbandry with reference to the particular problems involved and the nature of agricultural and veterinary organization that may be in existence or contemplated.*

We also need to keep in mind that with the advent of the Reforms in 1920, the direction of agriculture and veterinary activities passed from central government and imperial department to the provincial governments and provincial departments. The former, however, retained its responsibility for the promotion of research of all India importance as well as control over the research institutions at PUSA (Bihar) and Muktesar and continued to develop them within the limit of its means. Soon thereafter the need for some coordination of efforts began to be felt and Royal Commission on Agriculture (1927–'28) recommended the formation of Central Council of Agricultural Research to discover the way out of this difficulty. Following this recommendation, when Imperial Council of Agricultural Research was formed, the editorial responsibility of *The Agricultural Journal of India* was also transferred to new editors from ICAR, from the Agricultural Adviser to Government of India who had published it for the first 24 volumes.[17] Royal Commission on Agriculture had also recommended that the Board of Agriculture in India be reorganized into two wings, namely agriculture and animal husbandry.

The division of labour question from the point of view of which department was better endowed to engage with the cattle breeding operations was taken up for discussions again at the second meeting of veterinary officers in India held at Calcutta in 1923. Introducing the subject for the discussion, the chairman stated:

> I have referred to the subject in my opening address. I have nothing to add except to say that I shall be glad to welcome any suggestions dealing with the professional aspect, which we can embody in a resolution. I believe that in order to stimulate private enterprise, which is recognized to be the first step in any efforts destined to improve the cattle breeding industry, we should endeavor to form cattle breeding associations for the various breeds with the idea of eventually creating stud books.[18]

From the wording of the resolution that was passed at the end of discussion, it emerges that veterinary officers considered that the question of cattle breeding needed some advice from some central authority and veterinary officers should get adequate representation when such a central body/ provincial bodies get established:

> That this conference, strongly favours the establishment of Cattle Boards, Central and Provincial, and considers essential that the veterinary profession should be adequately represented on these bodies.

With the appointment of Arthur Olver as an imperial expert on animal husbandry in the year 1930, this question of which is the right agency for cattle breeding and animal husbandry resurfaced. At the first meeting of Animal Husbandry Wing in 1933, in a note written on the subject 'the organization necessary for the future development of animal husbandry in India,' Olver submitted figures to show how inadequate was the provision made by provincial governments for the development of animal husbandry by government agency. He had underlined that the figures were "only approximate because exact figures were not given in all cases of the amounts allotted for animal husbandry work from agricultural budgets". At the first meeting of animal husbandry wing, members passed a resolution that "The Board resolved that three permanent subcommittees of the Board of Agriculture, each consisting of four members, should be formed under three heads, viz. (a) animal nutrition, (b) cattle breeding and (c) dairying to deal with these branches of animal husbandry work". It was further decided that the chairman should fix the personnel of these committees in consultation with Colonel Olver.

Olver once again reconsidered the subject and wrote a similar note for the second meeting of Animal Husbandry Wing held at Madras from 14 to 16 December 1936 and compiled a table giving exact amounts which were allotted for animal husbandry wok during the year 1935–'36. From the data, Olver (1938: 247) arrived at two inferences, "i) allotment of funds for animal husbandry as a whole is still very inadequate and disproportionately small compared with what is spent on plant husbandry, and ii) there is still, in most provinces nothing which can be described as comprehensive animal husbandry organization, such as exists in other countries, with staff and facilities for dealing with all aspects of animal husbandry under the unified control of a specialist in such work".

Olver (1938: 247) reported that "only in the Punjab and North West Frontier Province is animal husbandry work of all kinds dealt with by one department, viz. the Provincial Veterinary Department under the unified command of a specialist". He also put forward the recommendation by Government of India to provincial governments that they should consider the formation of Provincial Animal Husbandry Departments. This proposal had arisen in situation where in many provinces cattle breeding and animal husbandry work was being carried out by provincial department of agriculture and in some other provinces by provincial Department of Veterinary and "unfortunately a fact that these directors do not always see eye to eye, particularly in the matter of organization of systematic breeding control," which in Olver's views, "is essential for adequate progress in the development of animal husbandry in India".

Olver (1938: 248) visualized a situation wherein "more specialists in dairying, sheep and goats and poultry and for marketing of animal products" were available and wherein "the Director of Animal Husbandry, should have at his disposal, funds and facilities more in conformity with the greater economic importance of livestock and livestock products in comparison with crops" than what was the prevailing case at that time. From the figures that he examined, Olver (1938: 248) realized that "total expenditure on animal husbandry work of all kinds, including disease control and the treatment of sick and veterinary education, is less than half the amount spent on plant husbandry,[19] though it is generally admitted that in India livestock are of special importance because they are used for cultivation of the land and milk is of special value because the diet of the people is predominantly vegetarian". Olver (1938: 248) was also at pains to point out that the average allotment for livestock improvement in the province where it was controlled by Provincial Agricultural Department was approximately 7.57% in the year 1935–'36, which was even less than 8.7% that he had cited in his note, 'The organization necessary for the future development of animal husbandry in India,' prepared for the first meeting of animal husbandry wing in the year 1933.

Even as he was putting forth the Government of India recommendation that envisioned provincial governments forming Provincial Animal Husbandry Departments at the second meeting of animal husbandry wing, Olver (1938: 55) was aware that he was again attempting to raise what was by then become a very controversial subject. However, "from his experience since taking up the present appointment, he was convinced that it was essential in every province or State to build up, from such veterinary and livestock personnel as existed, Provincial and State Animal Husbandry organizations, under the unified control of suitably trained specialists, who had no countervailing interests to consider". His suggestion came under sharp criticism from E J Bruen (1938: 56), who asked, "Why if veterinarian was the right person to control dairy does the Army Veterinary Corps not control the Military Dairy Farms?" Bruen wished to ensure that all branches get a fair chance in the new organization and was not willing to let go off the control over the 'cattle breeding' operations from officers from Department of Agriculture, as was the case in Bombay. Ware (1938: 56) wondered if this subject was going to become "a hardy annual" and in order to obtain agreement suggested the following three points for consideration:

> i) That the present staffs for livestock improvement work were inadequate for the development required.
>
> ii) That all possible use should be made of all suitably trained staff available in provinces and State.
>
> iii) That it was recognized that administrative questions might prevent the same organization being set up in each province but the principle to be adopted was that all staff employed on the development of livestock industries should be subject to unified control.

The chairman of the Animal Husbandry Wing (1938: 57) was of the opinion that "the expression 'unified command' in the last sentence was too vague and if intended to be precise was objectionable". The proceedings after giving a summary of views expressed and concluded that "the debate showed there was no agreement as to what department should exercise control". Even when the chairman put forward a suggestion that "instead of amalgamating existing Provincial Veterinary and Livestock Personnel into Provincial Animal Husbandry Departments, three departments should eventually be formed in each province or State, viz. agriculture, veterinary and animal husbandry". However, even this proposal was not acceptable to some members and the chairman said that "as the meeting was not a representative one and there was no agreement, he would not put any formal resolution".

The matter was again raised at the second cattle conference held at Shimla on 25 and 26 May 1937, when F. Ware (1937: 61) introduced the subject in the absence of Colonel Arthur Olver, who he states, "for the last seven years (1930–1937), has been doing his utmost and working extremely hard for the improvement of India's livestock industries" and who "in his numerous articles and notes expressed the opinion that what is required in this country more than anything else is a self-contained organization for animal industry". Suggesting relatively larger "use of the executive veterinary staff, where it is not used at present," Ware puts forth some data to argue out his case. He stated that in Madras the number of people in the Agricultural Department employed in livestock work was not more than 12; while on the other hand, Madras had a total of 297 in Veterinary Department, none of whom were used in livestock work! Similarly, in Bombay, the Agricultural Department had 21 on the staff who were doing any real livestock improvement work,[20] while there were 152 in the Veterinary Department and a number of them could be employed on livestock improvement work. In Bengal, there were just 6 men who were doing livestock improvement work, while the use might be made of a majority of 158 veterinary men. Ware did not present the figures on number of people respectively in Agricultural and Veterinary Department, but stated that the situation was of the same order with the only exception of Punjab, where 11 persons from Agricultural Department were doing livestock improvement work, but their work was augmented by 441 veterinary men, most of whom were employed for livestock improvement work. Ware had also referred to a curious situation where in the hurdle in making use of the executive staff in some province was that those men were employed under District Boards and were not controlled in ordinary way by the Director of the Department. Referring to a recommendation by the Royal Commission on Agriculture that had expected to define the duties of these men in such a way that ordinary dispensary work carried out by them got charged on local funds, whereas contagious disease work got charged on provincial funds, Ware suggested that the time was ripe to consider employing these men not only for disease control but also on livestock improvement work.

This prolonged discussion and debate on who shall control the cattle breeding operations; officers of Veterinary Departments or Agriculture Departments, suggests that 'technocracy' had emerged gradually to the fore by mid-1930s, whereby both veterinary side experts who dealt exclusively on the 'disease control' side and other livestock experts that looked after cattle breeding work, etc. guarded their respective positions with ardent zeal. The subject was very controversial, and the consensus was clearly lacking at all the previous occasions when officials held deliberations, so much so that W. Burns (1937: 63) prefaced his comment at the second cattle conference saying that discussing this subject he is reminded of the notice, "Electric Power Station, 10,000 Watts, High Tension, Keep Off". Burns then went on to present the organization structure of the US Department of Agriculture, where the department was split up into divisions; Bureau of Agricultural Economics, Bureau of Agricultural Engineering, Bureau of Animal Industry, Bureau of Biological Survey, Bureau of Chemistry and Soils, Bureau of Dairy Industry, Bureau of Entomology and Bureau of Plant Industry etc. Similarly in Canada, there were branches called Farms Branch, Dairy and Cold Storage Branch, Health of Animals Branch, Livestock Branch, Seed Branch, Entomology Branch, Fruit Branch, Agricultural Economics Branch, Publicity and Extension Branch. In Canada, all officials in the Health of Animals Branch were veterinary officers, while in Livestock Branch and Animal Husbandry Division of Experimental Farms Branch, the officials were agriculturists. By describing the organization structures in the USA, Canada and South Africa, W. Burns underlined a common element that various activities were entrusted to one very strong and powerful department, while in British India, the discussions on organizational aspects of animal husbandry were plagued by tenor of separatism, with an insistence on a separate budget, on separate personnel and a separate appeal to the cultivator.

W. Burns had referred to "the understandable feeling" that arose during the four-year period (1933–1937) that there was a possibility of an important branch of agricultural activity being split off from the agricultural department and being added to veterinary department. He also told the participants at the second cattle conference that he had a conversation with F. Ware where both considered the matter objectively and reached an agreement that "the real solution for this type of difficulty, which will be chronic unless we attempt to get some kind of a reasoned solution, is not separatism but unification". Thus, W. Burns had suggested solution in the form of "a powerful unified department of agriculture in which there will be only a single appeal to government for funds and there will be unified appeal in the form of a propaganda to the cultivator" and wished to see cooperation amongst those who worked on the questions of "the peasant's crops, the peasant's cattle and with the peasant's economics".

Moving on from what insights do we get about the nature of the colonial State and its imperial ambitions when we engage with history of cattle and organized dairying in India, in next part of this book, we shall discuss whether policy makers of post-colonial nation-state paid more focused attention to the peasant's crops, the peasant's cattle and with the peasant's economics.

Notes

1. In 1878, De Laval invented and perfected the continuous cream separator and thereby enabled separation of large quantities of milk in short time. The diffusion of this technology was rapid, as it had made its way in India within a decade, however, the growth rate of cream separators was very limited and there used to be less than 4000 of them by the year 1937, when Wright visited India. See, Wright, N. C. (1937) *Report on the Development of Cattle and Dairy Industries in India*, Government of India Press, Shimla.
2. For a brief account of the early private entrepreneurs such as A. C. Stafford, A. M. Kohler, G. R. Reeves and Pestonji Edulji Polson, who established modern dairy plants and ran a thriving trade in milk and milk products in Kheda district, see Heredia, Ruth (1997) *The Amul India Story*, Tata Mcgraw-Hill, New Delhi, pp. 17–26.
3. This paragraph on the role played by Military Dairy farms draws upon the Evidence of Lieutenant Colonel A. S. Mariott before Royal Commission on Agriculture. See Royal Commission on Agriculture, Vol I, Part II, (Evidence of Officers Serving under Government of India), Government of India, Central Publication Branch, Calcutta pp. 338–349.
4. This figure of butter output shows that the milk production for commercial purpose had taken a root in the Western India in the year 1916 in a decisive way.
5. Aden is a port city of Yemen that was under the British India administration until 1937.
6. Reproduced in the Proceedings on page no 51–52.
7. A short excerpt from the contemporary journals *The Farmer* and *Stockbreeder* titled, 'Increasing the Milk Yield,' reproduced in *The Journal of Dairying and Dairy Farming in India*, Vol IV, Part 2, p. 127.
8. *Royal Commission on Agriculture*, Vol I, Part II, Evidence of Officers Serving under Government of India, Government of India, Central Publication Branch, Calcutta p. 25.
9. Ghare, B. K. (1928) 'Preface to the First Edition (July 1923)' in in Ghare, B. K. (1928) *Manual of Dairy Farming for Indian Students*, Macmillan, Bombay, p. vii.
10. Smith, William (December 1927) 'Introduction' in Ghare, B. K. (1928) *Manual of Dairy Farming for Indian Students*, Macmillan, Bombay, p. xviii.
11. Proceedings of the Second Meeting of Veterinary Officers in India, p. 41.
12. For details see, *Miscellaneous Bulletin of Imperial Council of Agricultural Research*, No 27, Delhi, Manager of Publication, 1939.

13. Annual Report of the Department of Agriculture, Bombay Province for the year 1939–'40, pp. 150–151.
14. Agriculture and Animal Husbandry in India for the year 1938–'39, p. 178.
15. *Miscellaneous Bulletin of Imperial Council of Agricultural Research*, No 22, Delhi, Manager of Publication, 1939.
16. Ensminger, E. S. (1956) *Research in Animal Husbandry*, Indian Council of Agricultural Research, Delhi, p. 07.
17. Editorial to *The Agricultural Journal of India*, Vol XXV, Part 1, pp. 01–02.
18. *Proceedings of the Second Meeting of Veterinary Officers in India held at Calcutta (26 February to 2 March 1923)*, Government Printing Press, Shimla. In his opening speech, the chairman sounded hesitant to allow this subject to be discussed. He reminded his colleagues that the same matter was discussion at the first meeting that took place in 1919 and underlined that "in reviewing the proceedings of the last Veterinary Conference stated that they were not prepared to reconsider the matter at present". Even as he had allowed the subject to be on agenda with a hint of hesitation, he very firmly reminded his colleagues that "administrative questions are to be dealt with at the meetings of the Board of Agriculture".
19. Needless to say that after independence also the relatively higher allocation for crop/ plant husbandry compared to animal husbandry remained the trend in budget allocations. Off late, Livestock Development work has not been allotted funds proportionate to the share it contributed in Agricultural GDP.
20. This remark came under criticism from Mr. W. J. Jenkins who corrected Ware's data, by stating that out of the 260 people in the department, about half were dealing with livestock work in the district in addition to their agricultural duties. He felt it was absolutely necessary to point this out and to nail down the point for ever that "it is not only the staff of the Livestock Section of Agricultural Department who are doing livestock improvement work but a much greater proportion of the staff of the Agricultural Departments".

Chapter 7
Chasing Productivity Dreams: Intensification in Crop and Cattle Development in Post-colonial India

In the first part of this book, the narrative presented traces the shifting contours of crop-cattle interrelations and alongside that the shift in modes of agrarian production from pastoral to capital intensive and specialized practices. It has shown that how despite the advocacy for 'mixed farming', there had developed a model of milk production to satiate the urban demand for milk in the form of city milch stables. Even in peri-urban areas, there had arisen specialized dairy farm enterprises. Previous chapters have engaged with discussions amongst colonial experts and Indian nationalist intellectuals on the question of agricultural, veterinary and dairy sciences on issues such as cattle breeding policies and milk supply to emerging towns and cities. Most academic writings have looked at 'green' revolution in terms of technology transfer, miracle seeds, etc. but here in this chapter, the narrative engages with the dissenting voices who had talked about India's agricultural development with a critical lens on the logic of productivity and yields. These Gandhian thinkers were familiar with the writings of Sir Albert Howard and were rooting for organic farming and organic animal husbandry. Next two chapters will illustrate how the 'green' and 'white' revolution further accelerated yield and intensification-oriented planning in crop cultivation and cattle rearing. Alongside, next two chapters also trace emergence of government policies and programmes that unfolded in the realm of Agriculture and Cattle Development.

7.1 Gandhian Critique of Productivity

By early 1940s, Gandhi's call for 'Gau Seva' as an element within his overall rural development programme was gaining ground and in 1946, the Cow Conference held at Amritsar laid great stress upon the place cow held in India's rural economy.[1] An article penned by J. C. Kumarappa (1957: 12) in his magazine 'Gram Udyog Patrika' following this conference in November 1946 raised the critique of new variety of

H. Upadhyaya, *Critical Insights on Colonial Modes of Seeing Cattle in India (1850–1980)*, Asia in Transition 27, https://doi.org/10.1007/978-981-97-1560-2_7

long-staple cotton and the opening of Vanaspati "ghee" mills. An excerpt from his editorial below:

> From this broader approach any encouragement given to the cultivation of long staple cotton for mills tantamount to the destruction of the cow, as the seeds of long staple cotton are not available as cotton feed because of the fuzzy short staple cotton being left unlinted on the seed. Owing to this the bullocks are deprived of their protein diet. Our villagers are dependent on animals for the satisfactory working of their economy. The opening of Vanaspati "ghee" mills again cuts across this economy. It deprives people of a wholesome article of diet – vegetable oil and replaces it by indigestible, hydrogenated oil and sets up unfair competition with 'Telis'.

By May 1954, Kumarappa (1957: 17–19) was writing about the harm caused by tractor cultivation, even as there were only 8635 tractors in use throughout the country.[2] His arguments followed from his earlier critique of large industries and centralization that the young nation and its leaders appeared to have embraced. In February 1955, 'Gram Udyog Patrika' published an article by Om Prakash Sharma that reported about the experiments conducted by A. R. Khan, the agronomy specialist of the Indian Agricultural Research Institute, New Delhi, during 1942 to 1947. These experiments were conducted with the object of studying the effect of seed-bed preparation with alternative forms of tillage implements on the yield of wheat, by the treatment and comparison of bullock and tractor cultivation. These experiments had clearly showed that the deep ploughing of the tractor is harmful for wheat cultivation. Notwithstanding, these results of research by Khan, local tractor production in India began in the year 1961–'62 with output of 880 tractors per year and today India manufactures more than 2,65,000 tractors per year.

A pamphlet titled 'Why Gandhiji Fusses over Cow' dated 23 June 1946 elaborates arguments that 'Gau Seva Sangh' (A Gandhian organization devoted to serving the cow) put forward against mechanization of agriculture and usage of chemical fertilizers. The pamphlet cited opinions of Sir Albert Howard, former Economic Botanist to Government of India, from his book *An Agricultural Testament*.[3] Similarly, Dasgupta (1945: 20–26) narrated the story of Dr. Bratsch and his work at Marien-hohe in Germany under inspiration of Rudolf Steiner in order to point out that there was a harmony between soil, plant, humans and animals.[4] Having narrated the story of Dr. Bratsch and Sir Albert Howard, Dasgupta (1945: 33) pleads that it is time for India to have faith in her basic agricultural and animal husbandry practices and improve them in the light of the failure of the West. Dasgupta had hoped that India will work out "the methods of improving soil and animal health along old lines, enlightened by newer *scientific* knowledge of composting and bio-dynamics—things which already form the basis of Indian soil, plant and animal husbandry, without the cultivators knowing what those *sciences* were".

Even as Gandhian activists and thinkers were engaging with Sir Albert Howard's writings on organic agriculture, in 1943 a concerted effort was made to foster the use of chemical fertilizers by Indian farmers, with the setting up of Fertilisers and Chemicals Travancore Ltd and it started full-fledged production in 1947.[5] Within first five years of free-India, in 1952–53, the chemical fertilizer consumption stood at 65,685 tonnes. In next five years, it rose to 4,77,921 tonnes.[6] In 1960, Majumdar

(1960: 713–14) appealed to understand the role of cow dung as manure and lamented that "in discussions on investment criteria, the question of development of manure inputs in lieu of chemical fertilizers is not even considered, perhaps it is not regarded as sufficiently important to deserve study".[7] This was further accentuated in the years following the introduction of High Yielding Variety (HYV) seeds under the package programme. By 1968-'69, the fertilizer sales touched the level of 1.7 million tonnes. India's fertilizer industry was expanding rapidly and Brown (1971: 91) had predicted that by 1975 it could have a capacity to produce more than double the output it produced in 1968. This forecasting by Brown needs to be read along with a 1967 PL 480 report that had recorded tripling of fertilizer consumption in India within two years.[8]

7.2 The Bengal Famine, Partition and Emergence of the Food Policy

In contrast to Gandhians such as Kumarappa, Dasgupta and others stood those who pushed for reorganizing the agriculture to exorcise the ghost of Bengal Famine. By 1941, the per capita food grain availability figure had declined by 70 lbs, compared to 298 lbs per capita availability that was obtained in 1896.[9] On 6 April 1942, a conference of the representatives of the provincial governments in British India and Indian States issued a series of recommendations that culminated in the initiation of 'Grow More Food' as a slogan as well as a campaign.[10] The Grow More Food campaign advocated activities such as: (i) switch over from cash crops, mainly short-staple cotton, to food crops; (ii) intensive cultivation of cultivated lands through irrigation, improved seeds and manures and better farming practices; and (iii) extensive cultivation by bringing under plough current fallows, cultivable waste lands, etc. No targets for enhanced food production levels were set though, and hence, it was not possible until after 1947 to review the effectiveness and success of the campaign.[11] The wartime food crisis had forced a revision in the food policy in British India from *laissez faire* to highly centralized system of government procurement and controls with the establishment of a separate Department of Food in December 1942.[12]

The Bengal Famine of 1943 had suddenly aroused a national interest in the propaganda on increasing food grains yields. The logic of productivity was being repeated with identification of two defects that plagued Indian agricultural life; the absence of adequate irrigation facilities and absence of improved varieties of seeds capable of yielding a larger crop.[13] During the First Five Year Plan agriculture and irrigation received the greatest priority, just like the irrigation projects were prioritized in post-World War II reconstruction programmes. There were economic planners who argued to give a jolt to the Indian Agriculture in terms of its organization to face these challenges and one of them was Kolhatkar (1946: 81), who suggested increased mechanization of farming and didn't shy away from proposing that the easiest way of achieving this objective is to establish the industry of cattle-slaughter and meat-packing on a large scale.[14] The already precarious situation on the food front had been further endangered by the partition in 1947, when the wheat growing areas of West Punjab went to Pakistan. India received 82% of pre-partition population, but only 75% of cereal producing territory. In 1949, Agrarian Reforms Committee of Congress recommended complete and comprehensive land reforms as a pre-requisite in improvement of agricultural production and efficiency.[15]

In post-independence India, the Food Grain Policy Committee (September 1947) headed by Purushottamdas Thakurdas had arrived at the conclusion that the country was then in the short supply of food grains by at least 10 million tonnes. This led to 'Grow More Food Campaign' getting a five year extension, with a target of adding 4 million tonnes to food grain production. However, shortly thereafter arose another crisis when an acute shortage of cotton and jute surfaced owing to difficulties in imports from Pakistan. India had to resort to have an integrated programme for increasing production of food grain as well as cotton and jute from June 1950 onwards. Amidst such a fire fighting, a newly independent nation-state looked expectedly to USA and in 1949 held extensive dialogues on the possibilities of low-priced or barter trades of Indian minerals for wheat imports.[16] Wheat imports from the USA to India date back from 1946 during the post-war food crisis, but this received an upward swing after the enactment of PL 480.

The leaders of independent India were anxious to catch up with the developed world when it came to solve the problem of food grain production, agricultural research, etc. Their vision for research in agricultural and animal husbandry matter was inspired by what Soviet Russia and USA were doing. In his inaugural speech at the Seventh Meeting of Crop and Soil wing of Board of Agriculture and Animal Husbandry in India (1950: 05) Shri O. P. Ramaswami Reddiar, the then Premier of Madras asked:

> When are we going to have scientists like Burbank of California or Michurin of the Soviet Russia who working in the same field as Luther Burbank proved that under suitable conditions young hybrid seedling can be trained to develop any desired characteristics?...When are we going to have anything like the contribution of Soviet scientists in the matter of animal husbandry? When are we going to progress in matters like vernalisation of seeds and artificial insemination of cattle?[17]

In the above three rhetorical questions, one can get an idea of what was going through the minds of nationalist leaders who were eager to catch up with the Western, 'developed' world. Development was not merely a word indicating a material process; it had turned out to be an ideology.

At the onset of the First Five Year Plan, in 1951, the then Minister of Agriculture, Dr. Rajendra Prasad had described the food situation in India as "very grave" because the grain stock under central government had hit rock bottom, with an ominous possibility of breakdown of rationing in different provinces of the young nation state. B. L. Jalan (1951: 01) in his short booklet describing the 'Food Problem in India' reported that of the world food deficit of 18 million tonnes, India's deficit is 4.5 million tonnes, i.e. 25%.[18] During the period 1947–1951, the government efforts were directed more towards procurement plans both from abroad and within the country and its internal distribution, rather than increasing the aggregate food production, reported Jalan. He mainly offered a two-pronged strategy to come out of this impasse: (i) increasing the extensive and intensive margins of cultivation, and (ii) promoting a more balanced diet involving the growth of more vegetables, fruits, fats and oils and fish, meat and eggs.

It had been a decade since the Grow More Food Campaigns started in 1942 and an enquiry committee under the chairmanship of Sri V. T. Krishnamachari was constituted in February 1952 to review working of the campaign. Between 1947 and 1951, Grow More Food Campaigns had led to an increase in domestic food production at the rate of 3 million tonne per year, but imports had remained steady, occasionally even increasing to reach the level of 4.8 million tonnes in 1951. The committee recommended that Grow More Food campaign shall be expanded to cover wider rural development plans. The committee spoke in the idiom of the dominant mode of development thinking of the era which was distinctly premised on Community Development lines.[19] Amongst the more important recommendations of the committee were: (i) setting up of a countrywide extension organization, (ii) acceleration of minor irrigation programme and (iii) the provision of adequate rural credit. Thus, in 1952 Grow More Food Campaigns staff was merged with newly initiated Community Development Programme.[20]

Somewhat like Key Village Schemes set up for cattle development in post-independence India, Community Development Programme (1952–1960) recruited 50,000 workers across 500 blocks, each consisting of 100 rural villages. However, within a year, the US Technical Cooperation Mission that reviewed the programmes, expressed doubts on the approach and reliance on the then available technology. The mission in its report called for an expanded programme for training agricultural scientists and research.[21] In 1955, a Joint Indo-American team on Agricultural Research and Education advised for the need of creating agricultural education, research and extension service network in India on the lines of the US land grant universities.[22]

7.3 India's Agricultural Planning Under Influence of American Aid

Paul G. Hoffman, the then head of Ford Foundation, visited India in August 1951 at a time when technical cooperation from the US Department of Agriculture had begun in a small way. On his return, Hoffman hired a young extension specialist, Douglas Ensminger to head Foundation's soon to be opened Delhi office and by December 1951, Ford Foundation committed to extend $ 2.225 million financial assistance to Government of India. By December 1951, the Rockefeller Foundation had also committed to send a study team to India comprising experts who were involved in the Rockefeller Foundation's work in Mexico. Harrar, Mangelsdorf and Weaver wrote a document titled 'Notes on Indian Agriculture' (1952) and argued for an aid programme weaving themes of overpopulation, hunger, political instability and threat of communism.[23] This 22-page note outlined the future support on three aspects: (i) plant breeding work on rice and wheat, (ii) organizing agricultural education network in India on the lines of the US land grant universities and (iii) participation in agricultural extension activities within community development programmes. After four years of negotiations between Indian officials and foundation experts, differences were ironed out by late 1955 and in April 1956 Indian government signed a formal agreement inviting the Rockefeller Foundation to support with $ 1.38 million grant for the post-graduate school at Indian Agricultural Research Institute, New Delhi, and research work to raise yields of maize, sorghum and millets.

While agriculture and irrigation had received the highest priority in the First Five Year Plan (1951–'56), with an aim to add 7.6 million tonnes to the annual production of 54 million tonnes in the base year (1949–'50). Thus, the food grain production target was 61.6 million and the production level attained in 1953–'54 was 70 million tonnes and at the end of the first plan period was 65 million tonnes, thereby making India almost appear within the sight of solution to food problem. Now, controls were withdrawn, and food imports came down to less than a million tonnes in 1954 and 1955. In 1956, with the enactment of PL 480, there arose a situation where India could procure large amounts of wheat and rice from the USA. In August 1956, India signed an agreement to import 3.1 million tonnes of wheat and 0.19 million tonnes of rice over the next three years. Meanwhile prices of food grain had begun to show an upward trend. An alarmed government appointed the Food Grain Inquiry Committee headed by Ashok Mehta and S. R. Sen in 1957. The committee members considered ability to attain Second Plan target highly doubtful and ended up recommending in the short run to ensure steady inflow of food imports. Then with a drought in 1957 the food grain production took a knock plummeting output by 10%, to touch the level of 62.5 million tonnes.[24]

There was a strong conservative criticism of plans and dissidents in Congress formed under Rajagopalachari's leadership a new political party called Swatantra. A committee was appointed in February 1958 to study the problem and recommend the necessary action. Before its report could get published in October 1958, the situation had gone from bad to worse. The report of this committee, which came to be known as Nalagarh committee, identified low levels of agricultural production, low morale and inefficiency of administrative staff as reasons for the sad state of affairs. The report was probably far more visionary and suggested ways to hold and increase technical agricultural manpower. It had also argued to integrate at the state level all the ministries and departments dealings with programmes for development of agriculture, irrigation, land reclamation, soil conservation, animal husbandry, veterinary science, agricultural markets, forestry, cooperation, agricultural education, research and extension into one single ministry.

George Rosen (1985: 74) recollects how the Ford Foundation-sponsored team of experts was constituted:

> Douglas Ensminger, the head of Ford Foundation in India, was aware of these agricultural problems, and in 1958 traveled around India himself to see what was happening at the farm level. He was especially struck by the peasants' failure to adopt improved production practices, and reported this to Nehru and Sir V. T. Krishnamachari. He suggested that it

might be useful to have a team of foreign experts work with an Indian group to see what was happening in Indian agriculture and then make recommendations.

A team of American agricultural experts headed by Sherman, then chief economist of USDA's research service, was put together to study India's food problem. The team visited India for three months in early 1959, travelled around the vast country for 50–60 days in an Indian summer and wrote what became known as Food Crisis Report, published by Indian government in April 1959. The report uses a language that needs to be analysed by making a serious effort to read between the lines. At multiple places, it tries to give an impression as if the emphasis was on decentralized options, while at other it has repeated the oft expressed views on 'surplus cattle'. It also asked Indian government to introduce heavier tax rates on cattle, confinement and castration of scrub bulls and incentive payments for marketing of cattle hides and bones. At the heart of the report, however, was a thinking that technical information and rigorous advocacy for intensive use of yield increasing technology and inputs could be transferred and applied in concentrated fashion in more responsive areas.

The recommendations of the report which got published almost immediately by Government of India, suggesting almost a complete endorsement, however, did invite criticism from India's leading economists and government officers including M. L. Dantwala, V. K. R. V. Rao, Pitambar Pant as well as Daniel and Alice Thorner. Critics exposed flawed statistics that constructed a panic scenario in the face of secular growth in food grain production and argued that the analysis of the problem was incorrect and sought to rationalize shift in plan priorities from Industrial Development back to Agriculture.[25] The report had also foreclosed any hopeful thinking on cooperative agriculture and reoriented agricultural strategy from social reforms to higher levels of technologism.[26]

The next year Sherman Johnson and colleagues were invited to return to India to work with Indian experts in developing a specific action programme to implement the recommendations of the Food Crisis Report. In June 1960, a ten-point Intensive Agricultural Districts Programme, promoting a "package" approach to increase crop yields by intensive cultivation, was launched with the Ford Foundation contributing $ 10.5 million, about one-third of the cost over the next five years. The districts selected to implement IADP were mainly in wheat and rice growing areas, with adequate irrigation and assured rainfall and with minimum hazard of monsoon floods, except just one district in dryland millet growing area. An organizational framework to replicate the 'green' revolution that was seen through Dr. Norman Borlaug's work in the Rockefeller Foundation-funded Mexican Agricultural Programme was being laid out with the initiation of IADP in India.

IADP had provided empirical support for thinking that the scarce resources—fertilizer and water-responsive new seed varieties, fertilizers, pesticides, trained manpower and intensive programme effort—concentrated in areas with assured water and more responsive land resulted in higher output than other methods of allocation. In 1964 and 1965, Indian government decided to expand the scope of intensification strategy with concerted efforts on diffusion of high yielding varieties seeds under IAAP (140 districts) and HYVP (32 million acres of irrigated land in 250 districts) leading to an official announcement of New Agriculture Strategy under the leadership of then Minister of Agriculture C. Subramanian in 1966. This was now possible since new hybrid varieties of wheat and paddy exotically developed in Mexico and Philippines respectively and new locally developed hybrid maize, jowar and bajra had recorded 25–200% higher yields per unit of land or fertilizer compared to local non-hybrid varieties.[27] However, even as an expansion in scope to replicate the principle was taking place, the Ford Foundation after the year 1965-'66, decided to concentrate its budget and technical assistance in five IADP districts, viz. Ludhiana, Raipur, Sambalpur, Thanjavur and West Godavari, and these districts came to act as spearhead spots where to try a new technological innovation.

7.4 Settling the Nomad

After India won independence, the members of Animal Husbandry Wing of ICAR, started to think about 'the role of professional and nomadic cattle breeder in the rural economy of India,' but the mode of thinking had remained 'paternalistic' that could only think of sedentarization as the only alternative. Report of the Cooperative Planning Committee (1946: 44) had taken note that there were a few nomadic tribes like Rabaris of Gujarat, who own and breed cows and buffaloes and Odes of Punjab, who own and breed mainly sheep. With a view to providing these people with facilities for improving their animals, the committee had recommended (1946: 44) that "they should be organized into cooperative societies, inducement being provided by the State, or local authorities by reserving grazing areas for such societies". Towards this purpose, the committee had recommended that "surveys should be made with a view to finding out whether grazing land exists which could be allotted to these societies, and government farms should supply pedigree bulls and rams to them".[28]

Acting on this recommendation, the Government of Bombay had already launched such a rehabilitation scheme for nomadic cattle breeders. In the annual report of the Department of Animal Husbandry and Veterinary Science of Bombay Government for the year 1950–'51, one also comes across a similar thinking that had informed a scheme for these nomadic communities initiated by the Native State of Baroda a decade earlier:

> The scheme was launched by the Government of Bombay in the year 1948–'49 with a view to settling on land members of the Rabari and Bharwad communities of North Gujarat and converting them into cultivator-cum-cattle breeders by granting them sufficient land for grazing their cattle and for raising crops for maintenance of their families. A similar scheme had been started by the former Baroda Government in the Mehsana district in the year 1939–'40, and on the merger of the Baroda state with Bombay, the two Rabari colonies, one at Pratapnagar and other at Bellasar, and the 24 settlements started under that scheme were taken over by the department. During the year under report the scheme run by the Bombay Government and the old one started by the Baroda Government were amalgamated and reorganized. The number of families living at the two colonies and 24 settlements was 406 with a total population of about 2181 souls. The total number of cattle maintained by these families was 5219 and the number of breeding bulls 17. The area under crops was 5827

bighas and 13 Vasas and the approximate yields were 23,308 mds of food grains and 72,838 mds of fodder.[29]

At the eighth meeting of Animal Husbandry Wing of Board of Agriculture and Animal Husbandry in India that was held at Mysore during 17–22 February 1949 (1951: 28), experts discussed 'the steps that may be taken to make professional and nomadic cattle breeder play a greater part in the improvement of cattle'. The committee that had considered the subject and submitted its report prior to the meeting felt that "they require a lot of assistance, direction and education to enable them to play a greater role in the rural economy of the country".[30] The committee (1951: 28) had suggested the following:

(a) Encouraging the breeder to settle on land where feasible.
(b) Providing education facilities for their children as well as adults, the education imparted having a bias towards animal husbandry.
(c) Encouraging better animal husbandry practices by providing pedigree bulls free of cost and castration of scrub animals, if necessary by legislation.
(d) Improving the methods of feeding by facilitating the supply of concentrates and mineral mixtures, effecting improvements of the grazing areas by research especially to discover strains of hardy grasses in the over-grazed areas of India, introducing better methods of hay making, grass preservation, silage and utilization of tree leaves, etc. In addition arrangements for the permanent supply of drinking water are also necessary.
(e) Providing efficient medical and veterinary assistance.
(f) Helping them to dispose of their livestock and products like ghee, butter and milk and purchase their daily requirements on cooperative basis and provide cheap credit facilities.

However, it is pertinent to note that at the meeting of animal husbandry wing, there were members who raised questions on certain recommendations and especially those that had voiced a clear bias against the 'mobility' of professional and nomadic cattle breeders. For example, voicing his views against the recommendation that attempted to prohibit 'pastoral mobility', Zal R. Kothawala (1951: 29) felt "referring to the first recommendation in connection with the settlement of breeders on land, that such a measure would throw many natural grazing, which they had at present, (if those) were taken away from them, they would begin to suffer". He stated that such an experiment had been tried in certain parts of Bombay Presidency, but the results were hardly encouraging. He pleaded his colleagues to recognize that "there are three types of such breeders—those in forest areas, those not far from inhabited area and those in the desert area," and believed that "while making the recommendation, the committee had perhaps in view only one type of nomadic breeders, namely those in forest areas such as Central Provinces". About the nomadic breeders in the desert area, Kothawala had an opinion that they "were by far the most important one" and "if these people were tied to a particular region that would kill their profession".

Similar opinion was expressed in a note prepared for the meeting by B. B. Jamaspyina (1951: 143–144), when he stated that although Rabaris and Bharwads have become some sort of a 'nuisance,' 'burden' or 'pest' upon the land due to diminishing grazing land but increasing stock, "the quality of the cattle of the famous

Kankrej breed which he possesses is so high that he cannot be asked curtly to reduce his stock" since doing so would be a "*national calamity*" and a "very fine breed will be partly lost (emphasis in original). Jamaspyina recommended that "a practical and useful scheme should be formulated by which he would thrive, and thus keep up the fertility of the soil".

When the first recommendation came under such criticism by Kothawala, who had served first as an officer in charge of Dairying at Bombay Municipality and later as Imperial Dairy Expert, the committee chairman Hurley (1951: 29) clarified the matter informing that "the first recommendation was based on majority opinion only," and that the "recommendation was elastic and could be introduced only where feasible". Hurley said:

> The idea of that recommendation was to prevent those people from wandering about to the extent possible, though it was difficult to formulate a uniform policy for this purpose as the habits and methods of these breeders varied a lot in different areas. What the Livestock Committee had in mind was that those breeders might be made to stay in one place for somewhat longer periods which would help in improving the land.

The chairman of Animal Husbandry Wing, Datar Singh (1951: 30) felt that "once the breeders were settled on land they would no longer remain as nomadic breeders as such, but would ultimately become cultivators". He also reported that "the nomadic breeders from Gujarat had represented several times that most of their land (sic) had come under cultivation, and owing to constant fighting between them and the cultivators, those breeders had become something of a menace. They represented that they had now no grazing areas and as they did not know where to go, they wanted that some land should be given to enable them to settle themselves."[31]

Here we can see how the contestation between a cultivator and a nomadic herder was being interpreted by officials and experts as the latter being considered a 'constant menace,' whereas the action of the former is always looked upon more sympathetically as arising from the desirable goal of bringing more land under cultivation with 'Grow More Food' thinking. Thus, while the discussion on the committee's recommendations witnessed members taking different positions, thereby showing that all of them did not view 'mobility' as a 'menace' and 'settling down' nomads as a desirable option, the recommendation that the chairman (1951: 32) proposed was:

> Wherever it was found necessary to take away the best grazing areas for cultivation, due consideration should be paid to the need for settling these nomadic breeders in those areas on the land keeping in view the need to ensure that they remained primarily as cattle breeders.

By adopting this recommendation, the board had clearly displayed that in those early years after India's independence, to bring more land under cultivation and to increase food production had been accorded priority compared to preserving grazing land and pastoral mobility. If necessary, even the best grazing areas could be alienated from the customary and traditionally established usage and land use classification. Thus, the term 'protected areas' was not to be attached to "even the best grazing areas" in an expert imagination. Similar view is expressed by Whyte (1961: 353), when writing about 'Nomadic and Migratory Grazing,' he asked:

> How is the improver of the nomadic way of life and its characteristic types of domestic livestock to approach the problem of the amelioration of semi-arid environment or in some cases the elimination, total or partial, of this type of animal husbandry from the land use system?[32]

Whyte (1961: 354) thought that the best way to answer this question and to provide rest to the grazing ranges at the receiving end of the intensive grazing for excessively long period of year, under the existing social and economic conditions is to hold back nomadic and migratory flocks and herds, for say one month longer in the cultivated land and bring them back to these lands one month earlier. Whyte (1961: 355) at the onset of green revolution sort of intensification was probably a lone voice stressing upon an intensification and integration of a different kind, since he observed "there is practically no integration of crop and livestock husbandry in the cultivated lands—there still persists the fundamental distinction between cultivator and shepherd that may have originated at the dawn of agriculture". Whyte recommended that "the main objective in these areas must be the fullest integration of crops and livestock, and thus to bring the cultivated land which needs it so much, the animal fertility, the droppings of millions of livestock which are at present wasted on the desert sands".

Reporting on the efforts by the post-independent nation-state to rehabilitate the nomadic cattle breeder, Whyte (1961: 367) wrote:

> A Nomadic Cattle Breeders' Scheme in the Second Five Year Plan was taken up in former Bombay state, Rajasthan and Uttar Pradesh. While the Bombay scheme provides for organizing cattle breeders into cooperative farming-cum-breeding societies and settling them on waste lands, the scheme in Rajasthan plans improvement in the condition of breeders through grasslands development, improving watering facilities, offering the subsidy for rearing of selected cattle, providing breeding bulls, improving marketing conditions and providing veterinary aid. The Uttar Pradesh scheme envisages improvement of grasslands, free veterinary aid and provision of breeding bulls. Already in Bombay, over 66 cooperative breeding-cum-farming societies have been organized and 1215 breeders settled on nearly 21,300 acres.

Thus, one can witness in these early discussions that policy makers and experts did think about what the impact of their interventions in the field of agricultural development and increased crop production on the landless nomadic and pastoral populations would be. Yes, one can argue that there was a continuity in their thinking where they like their predecessors did display a sedentary bias and thought of making efforts to 'rehabilitate' nomads, as if they were a 'vanishing tribe'. However, one also needs to appreciate that these first-generation planners in post-colonial India were aware about the implication of their policies for landless nomadic and pastoral populations and at least there were some acknowledgements of impacts, even though accompanied by paternalism.

7.5 Key Village Schemes for Cattle Development (1951–1965)

The first organized efforts to develop cattle in rural areas on an effective scale in post-independence India began with the introduction of Key Village Schemes during the First Five Year Plan (1951–'52 to 1955–'56). The thinking on these lines had started amongst animal husbandry experts of Indian Council of Agricultural Research even prior to the First Five Year Plan period. At the eighth meeting of Animal Husbandry Wing of the Board of Agriculture and Animal Husbandry in India, Sardar Datar Singh, the chairman (1949: 04) spoke thus in his inaugural speech:

> I wish to refer to one subject, which I consider to be of far reaching importance, viz. the establishment of Key Villages for improvement of cattle in all the breeding tracts of India. The more I think about the complicated problem of Livestock Development, the more is the conviction forced upon me that such a scheme is the *sine qua non* of success in the economic exploitation of cattle in India. As you are aware these Key Villages are intended to be compact areas where concentrated work can be carried out under proper control and supervision and continued for years. According to rough calculation, a network of 625 Key Villages spread over the whole country will serve the purpose of producing superior livestock for improvement in the country at large. These Key Villages are meant to be real laboratories where the various measures proposed for animal husbandry development can be introduced and worked and where the results can be properly watched. It would not be practicable to take up the whole country in the scheme, but we hope that the improvements will radiate from the Key villages and spread into the country on an ever-widening scale. By adopting this method, we should be able to cover the whole country with improved cattle in the course of 15 to 20 years which will give more milk and better bullocks.

Back then it was being imagined that a period of 15–20 years was required to make the scheme reach the entire length and breadth of the country and the purpose was to give those villages 'more milk and better bullocks' by making the cattle breeding policy rest squarely on aiming for 'dual purpose animal'. Dairy development received total outlay worth Rs. 7.8 crore in the First Five Year Plan and in 1947, M. L. Sikka initiated Greater Calcutta Milk Scheme in West Bengal. In the Second Plan, dairy development outlay got enhanced to Rs. 19 crore, and in the Third Plan, it was further enhanced to Rs. 39 crore.[33]

Key Village Scheme (KVS) remained the most important animal husbandry development programme for the period covering the first three five year plans. The Key Village Scheme (initiated in 1951–52) aimed at establishing throughout the country a number of centres, each consisting of three or four villages, where breeding was supposed to be confined to a few superior bulls of known pedigree. During the first two years of the First Five Year Plan, 222 key villages and 96 Artificial Insemination Centres were established, and by the end of 1953–54, the number had increased to 345 key villages and 112 Artificial Insemination Centres, reported the reference annual compiled by Research and Reference Division, Ministry of Information and Broadcasting, Government of India (1955: 176). By the end of the First Five Year Plan, 545 key villages and 146 Artificial Insemination Centres were established.

This scheme also proposed to remove or castrate all other bulls. This scheme had proposed to relocate all the non-de-script bulls to *Gosadans* (home for unproductive bovines, generally established in forest area away from cultivated land), from villages which have 500 or more cows. The reference annual (1954: 160) reported that "of the 160 *Gosadans* to be established under the First Five Year Plan, 18 were sanctioned during the year 1952–53". The reference annual published next year (1955: 176) noted on a sombre note that, "*Gosadan* scheme has not made satisfactory progress; only 10 *Gosadans* had been established by the end of 1953–54". Eventually, it was realized that this component of the scheme had met with a complete failure as the state governments failed to provide the funds necessary for the setting up and maintenance of *Gosadans*.

Key Village Schemes also recognized artificial insemination as a key to the success of cattle improvement and 138 AI centres were opened in 522 key villages up to the year 1954–'55, with a target to take it to 150 centres in 600 key villages by the end of the First Five Year Plan. Detailing the KVS programme, Chakravarthy (1985: 26) states:

> The Key Village Scheme (KVS) embraces all aspects of cattle development including controlled breeding, better management and marketing, and adoption of improved animal husbandry practices through proper extension methods. It is a multifaceted scheme, but the basic objective is the multiplication of superior germ plasm from the established farms for improvement of cattle in other areas. Before the implementation of KVS on all India basis, artificial insemination (AI) techniques in cattle breeding had already been applied successfully. Therefore, AI became an integral part of the programme. The KVS centres were thus first established in the breeding tracts and were then extended into other parts with no recognized breeds, or in areas of non-de-script cattle. During the first part under this scheme, 146 Key Village Blocks (KVBs) were started, each block covering 10,000 breed able cows and she-buffaloes (i.e. serving only 1.9 percent of the breed able cows and she-buffaloes in India). Under the Second Five Year Plan (1956–'57 to 1960–'61), the scheme was expanded to 197 Blocks and 64 Urban AI centres.

Ensminger (1956: 19) points out that by the year 1954–'55, artificial insemination had become increasingly popular in the key villages and during that year 67,390 cows and 29,743 buffaloes were artificially inseminated.[34] As Ensminger (1956: 25) narrates, Key Village Scheme attained considerable success during the First Five Year Plan, and in the Second Five Year Plan, it was to be expanded on a much larger scale, with establishment of 245 AI centres covering a much larger area and a larger number of animals (5000 cows per centre instead of 2000) than was done during the First Five Year Plan. George (1988: 02) reports that by the end of Second Five Year Plan, about 1858 Key Village Blocks, 395 AI centres, 254 extension centres and 1900 veterinary dispensaries had been established.

In addition to its thrust on artificial insemination, KVS also paid some attention to addressing the fodder question and initiated Feed and Fodder Development Scheme from the year 1959 as one component in the programme. In February 1959, it was reported that the scheme had already been sanctioned in Assam, Bombay, Madras and Mysore, and it was in active consideration by other state governments. Whyte (1961: 03) reports that Madhya Pradesh had a similar scheme of its own and the Ministry scheme was approved by Punjab for implementation from 1959-'60. He adds that the scheme was expected to be taken up by Jammu and Kashmir, Himachal Pradesh and Manipur in the same year. Detailing the features of the scheme, Whyte (1961: 03–04) states, "This scheme which is subsidized by the Central Government on a 25 or 50% basis, incorporates the distribution of seed and planting material, the establishment of 10 acre pasture development plots, the development of pasture areas at State Livestock Farm, and subsidies for construction of village silos and for the supply of supplementary oilcakes and minerals". The scheme also involved appointment of Fodder Development Officer in the State Animal Husbandry Departments for the implementation and cost was to be shared on a 50: 50 basis between Centre and State. Sen (1957) provided an estimate that an increase in the fodder crop acreage of about 10% was necessary to provide an improved fodder basis, particularly of green fodder.

George and Nair (1990: 02) state that towards the end of the Second Plan nearly 600 KVS centres were functioning in the country covering an estimated 6 million cows and buffaloes, i.e. 10% of the total stock. Not only was the coverage limited, but according to an official review of the project in 1961, its contribution to the improvement in genetic quality and the productivity of the stock was also insignificant. It was envisaged to expand the scope of the improvement programmes and the area covered during the Third Plan period (1962–'61 to 1965–'66). While, during the first two plans, the scheme expenditure was shared between the Government of India and state governments, during Third Plan, the KVS was transferred to the State sector and direct financial assistance from the Government of India discontinued.[35] In a way, this shift in the pattern of financial allocation affected the capacity of different state governments to provide adequate inputs and services to livestock keepers.

Meanwhile, the official review of Key Village Schemes (KVS) in 1961 had perceived failure of the programme to make significant impact accompanied by the continued shortage of fluid milk supply in rapidly growing urban areas.[36] Large number of dairy plants set up during second and third plans had failed to collect sufficient quantities of milk and used to run much below capacity. This led government to target the already developed milk producing tracts in an intensive fashion. Another programme review report penned by Kaushal (1963) underlined the weaknesses of KVS by stating that, in view of budgetary provisions between the federal and state governments, the KVS has not been able to make any significant impact on the improvement of cattle breeds.[37] Particularly, the centres were located in dispersed manner all throughout the country but in many centres all the facilities envisaged in the KVS were not available and implemented. Each Key Village Block thus turned out to be a small island of organized activities surrounded by extensive areas of large cattle population on which the scheme had no effect. Therefore, the impact of the work done by KVS was said to get diffused by the uncontrolled breeding going on all around the countryside. George (1988: 02) points out that a major weakness of KVS policy was also its exclusive bias in favour of cattle development and its virtual neglect of buffaloes.

Crossbreeding with imported foreign bulls was mainly carried out in the hilly tracts and plantation estates around Shillong as well as certain southern regions such as Madras, Coimbatore, Nilgiris, etc. Ensminger (1956: 17) reports that with the total stoppage of the importation of foreign bulls, the above-mentioned areas were compelled to use such crossbred bulls as they can get hold of. However, by the early

1950s, even these crossbred bulls were becoming practically unobtainable, and there were insistent demands for importation of foreign bulls for crossbreeding purpose. Ensminger (1956: 17) thus reports that ICAR, therefore reviewed the position in all its aspects in 1953 and decided to investigate whether it was possible to reintroduce crossbreeding in the hilly and heavy rainfall areas of India, where the results of grading up with indigenous bulls were found to be slow and unsatisfactory.[38] Crossbreeding was introduced in late fifties on an experimental basis in hilly and heavy rainfall areas.[39] ICAR had drawn up around a dozen schemes in different provinces to carry out intensive work on cattle breeding by Jeresy semen airlifted from Central Artificial Insemination Station located at Hessarghatta, near Bangalore.

Thus, when one looks at the policy documents from the early years after independence, one can understand that while 'dual purpose' cattle breeding remained firmly a guiding principle, only for the hilly tracts and plantation areas where resources were better, a strategy was introduced to carry out crossbreeding in limited fashion.

7.6 From Cooperative Dairying for City Milk Supply to Milk Colonies (1946 to the Second Five Year Plan)

The seventh meeting of animal husbandry wing in 1946 (1951: 87) had discussed on the question of 'the Economics of Village Milk Production and the Improvement of the Marketing of Dairy Produce as to ensure a satisfactory price to the producer'. At the next meeting the replies received from various provinces and states suggested that the Governments of United Provinces, Central Provinces, Travancore and Cochin either had already established or had proposed to establish Dairy Cooperative Societies.

The eighth meeting of animal husbandry wing at Mysore from 17 to 22 February 1949 (1951: 42) recommended encouraging Cooperative Milk Societies to replace the then prevailing system of supply of milk produced in the heart of urban areas. The committee felt that the production side may be handled on cooperative lines and the processing and marketing by the Government Department concerned, by employing well-qualified dairy staff.[40] Very early on at this meeting, we witness J. S. Grewal speaking for the East Punjab and voicing his concerns that if this was done on a large scale it would mean depriving the villagers of their requirement of milk and milk products. Grewal had clearly warned that the entire exercise could end up being a rural–urban drain making milk 'the only fluid that flows upward'. Grewal preferred the strategy of establishing dairy colonies near and round about the cities and in East Punjab province convinced the government to set up 200 such colonies in 13 districts.[41]

The reference annual compiled by Research and Reference Division, Ministry of Information and Broadcasting, Government of India (1954:371) reported the following on Aarey Milk Colony in the chapter on Bombay state, contributed by Girija Shankar Bajpayi:

> The Aarey Milk Colony near Bombay maintained its progress and supplied pure pasteurized bottled milk to the people at reasonable prices. Today, it has 12 units with a cattle population of over 12,000; the daily processing of milk amounting to 3200 mds.

The reference annual compiled by Research and Reference Division, Ministry of Information and Broadcasting, Government of India (1955: 176–177) reported the following developments on the subject of Dairying:

> A number of states including Bombay, Madras, Madhya Pradesh, Uttar Pradesh and West Bengal, have made provisions for dairy schemes amounting to Rs. 7.81 crore in their plans. The Bombay Milk Supply Scheme at Aarey now has 14,526 heads of cattle and in conjunction with the Kaira District Co-operative Milk Producers' Union, Anand, supplies 3,500 mds of milk daily to one million consumers in Bombay city. Co-operative milk societies are also functioning in Poona, Dharwar and Ahmedabad. In Uttar Pradesh, the six co-operative unions organized for supplying milk to the towns of Lucknow, Allahabad, Banaras, Kanpur, Meerut and Haldwani distributed about 1.75 lakh maunds of milk in 1953-54, against 1.29 lakh maunds in 1950-51. In Madras state, 41 unions with the help of 465 co-operative societies catered to the needs of consumers in Madras city, Coimbatore, Tiruchirapalli, Madurai, Tirunelveli, etc. In West Bengal, the Government farm at Haringhatta supplies 300 mds of milk daily to Calcutta.
>
> In the original plan there was no provision for Central assistance for the dairy schemes. The position has now been reviewed, and Central assistance is now provided. The following proposals have been approved in this connection:
>
> (i) Expansion of the Calcutta Scheme by providing for the removal of 4000 milch cattle from Calcutta city to Haringhatta, and starting a 670-acre fodder farm at Kalyani at a cost of Rs. 71 lakh during the Plan period.
> (ii) Expansion of the Bombay Scheme by starting a second colony for housing 10,000 cattle on the lines of Aarey and starting a dry cattle and salvage farm at a cost of Rs 350 lakh during the Plan period.
> (iii) Removal of 2000 milch animals from the Delhi area to a colony near the city where a processing plant, dairy and dry stock farm will be set up. The cost of the scheme in the Plan period will be Rs. 47 lakh.

Five years later the same reference annual (1960: 257–258), reported the following developments on the dairy development schemes:

> During 1959–60, a provision of Rs. 275 lakhs was made as Central assistance for completing dairy development schemes undertaken previously and for starting new ones. Besides, an amount of Rs. 77.3 lakhs was provided for the Delhi milk scheme. The central dairy of Delhi Milk Scheme started functioning on a modest scale from 1st November 1959. The milk colony at Madhavram near Madras was started in November 1959 with 300 milch animals housed in one unit. The milk colony at Haringhatta near Calcutta added 2000 milch animals to its number thus raising the total to about 5000 animals. The Haringhatta milk plant enhanced the handling to about 1300 mds of milk daily. The dairy at Aarey Milk Colony is handling nearly 5000 mds of milk daily. The dairy plant installed by Guntur Co-operative Milk Union has been put into commission. Buildings have been constructed for the dairies at Chandigarh, Gaya, Trivandrum, Bhopal, Agartala and Coimbatore and arrangements have been made for the procurement and installments of equipment at these centres. Construction work has also been started for dairies at Patna, Jaipur, Hissar, Lucknow and preliminaries completed for starting construction work at Agra, Nellore, Cuttack and Srinagar.

> The Kaira Cooperative Milk Union, Anand, continued to make steady progress in the production of butter, milk powder and condensed milk. Work has been in progress on buildings for the Amritsar milk product factory. Preliminaries were completed for starting the construction of a similar factory at Rajkot and three rural creameries Baraui, Junagarh and Aligarh.
>
> A hostel with aid from New Zealand Government was completed and inaugurated at the Aarey Milk Colony. Another £ 800,000 received as aid from New Zealand under Colombo Plan was utilized for the Delhi Milk Scheme. An additional aid of £ 255,000 from that country is being utilized for securing equipment for Madras and Patna projects. The UNICEF made an additional contribution of $ 317,000 for the Rajkot and Ahmedabad projects. The UNICEF also made an allocation of one million dollars for Bombay's second dairy and $ 590,000 for the Bangalore dairy project.
>
> In plant training of technical personnel from the States is being conducted at the milk plants at Aarey, Haringhatta and Anand and five persons have been deputed for training abroad under various foreign programs.

Das (1960: 75–76) talks about the proposal that featured in Second Five Year Plan with an aim to expand the cooperative dairying in India stating, "under the Second Five Year Plan, it is proposed to organize 36 urban milk supply schemes, 12 cooperative creameries and 7 milk drying plants".[42] Milk producers' cooperatives were expected to be organized in villages to supply milk to the urban milk supply schemes, creameries and milk drying plants. It was also proposed to promote the distribution of toned milk in urban areas. Das had also pointed out that there were a few states that have made some progress in the sphere of cooperative milk supply. Das reported that in Bombay, there were 225 milk producers' societies and 11 milk producers' unions at the end of June 1958. Similarly, he states that there were 15 cooperative milk supply unions and 283 cooperative milk supply societies in Andhra Pradesh in 1957–'58; while in Madras, there were 20 cooperative milk supply unions and 621 cooperative milk supply societies and Uttar Pradesh had 9 cooperative milk unions and 258 milk supply societies in 1958. Das also alludes to similar cooperative milk supply societies and unions in some other states, including Kerala, Mysore and West Bengal.

This clearly suggests that much before 'replicating Anand pattern cooperatives' strategy that emerged following creation of the National Dairy Development Board in 1965, followed by launch of Operation Flood from 1970, the cooperative dairying had become a mainstay in India right in the late 1950s and there were several players in that institutional form.[43] The planners and policy makers were already stressing on extending state support to these initiatives. However, a close scrutiny of the dairy development and animal husbandry policies during the period between First to Third Five Year Plan suggests that cooperative dairying had suffered from lack of integration with city milk supplies, with the exception of KDCMPU.[44]

In the year 1966, the reference annual published by the Ministry of Information and Broadcasting (1966: 237–238) stated the following while reporting on the progress under Dairy Development:

> Dairy development programmes include establishment of city milk plants, colonization of cattle, establishment of milk products factories and rural creameries, rural dairy extension and training of technical personnel. Two new dairy plants were commissioned during the

year at Baroda and Bangalore, thereby raising the total number of organized milk plants to 30. In addition, 35 other milk schemes are in different stages of implementation. Of these, equipment for 12 dairy plants costing Rs. 81 lakh has been contracted under the Swedish credit programme, while equipment worth Rs. 22 lakh for another 6 dairies is being obtained under the Danish credit programme. The present average daily milk thorough-put of the organized dairies is 13 lakh litres compared to 10 lakh litres a day during 1964-65. The State Government of Punjab, West Bengal and Gujarat have issued milk control orders which have resulted in substantial increase in thorough-put of dairies in these states. *The cattle feed compounding factory at Anand, set up in 1964–65, produced about 100 tonnes of mixed feed a day.*

Two milk powder factories oriented to defence needs – one at Anand and the other at Mehsana – were commissioned during the year. Already two similar factories are in operation at Amritsar and Rajkot. In order to meet the emergent needs of defence forces in September 1965, arrangements were made of the daily supply of 23 tonnes of whole-milk powder of which 15 tonnes were through public/cooperative milk powder plants and 8 tonnes through private sector plants.

In addition to the creameries at Aligarh and Barauni, a new creamery was commissioned at Junagarh, thus completing the setting up of all the USAID-assisted creameries. The construction work of composite milk plants at Vijaywada with UNICEF assistance at Miraj under the Danish credit programme is nearing completion. The UNICEF will also supply dairy equipment worth Rs. 38.50 lakh for the Ludhiana composite milk plant. All the major item of dairy equipment excepting bottling lines, milk drying equipment and certain ancillary items such as stainless steel pipes and fittings are now produced within the country.

Training in dairying was imparted at the six centres functioning at Karnal, Bangalore, Allahabad, Anand, Aarey and Haringhatta. During the year, a diploma course in dairy engineering and a tutorial workshop for dairy teachers were also conducted. Nine persons were sent to Denmark for training in different branches of dairying under the FAO programme.

7.7 From Key Village to Intensive Cattle Development (1965 Onwards)

In 1964–'65, the Government of India approved a programme for the improvement of cattle in milk procurement area. Initially, Intensive Cattle Development Projects were started in 16 districts, each with its own project officer, who was responsible to the Director of Animal Husbandry in that state. By the year 1966–'67, there were 21 ICDP sites, which rose to 62 in the year 1972–'73 and it further went on to rise to 82, by the year 1975–'76.[45] The objective in each case was to increase the production of milk by one-third by the end of the Fourth Plan period. Giving us the background on why there was a shift from KVS to ICDP, Chakravarthy (1985: 34) states:

> It was realized during the Second Plan that KVS would not make significant impact on the improvement of stock on account of insufficient inputs. These problems led to the formulation and establishment of Intensive Cattle Development Projects (ICDP) as a part of special development programme during the latter half of the Third Plan. The Principal objective of the ICDP was to improve bovine breeds for higher milk production. Therefore the projects were located in the breeding tracts of indigenous breeds of cattle and buffaloes, and also in the milk sheds of the large dairy projects. The linking of the ICDP with the fluid milk marketing schemes and milk product manufacturing plants was considered essential

since cattle improvement programs had to function complimentary to milk production and marketing. The programme also included improved methods of breeding, provision of more feed and fodder production, disease control and adequate coverage for all round development of animals.

Even as these arguments for intensification of dairy development surfaced, Panse (1961) and Sukhatme (1962), those two stalwarts of Indian Agricultural Statistics, had argued against the diversion of resources to increase supplies of foods of animal origin, when a large part of Indian population remained undernourished due to insufficient cereal and vegetable intake. We need to know the context in which ICAR, Panse and Sukhatme opinions were offered. Was it in the context of the proposals for crossbreeding of cattle in India with European dairy breeds in order to increase milk production? Indeed, as Rao, Venkatsubramanian and Wit (1995: A 112) notes, ICAR appeared to have concurred with the recommendations by the Sarva Seva Sangh (A Gandhian Organization working on the issue of 'cow protection') to develop large non-de-script cattle into dual purpose animals combining draught and milk qualities in optimum manner in the year 1949, thereby generally agreeing to views put forward by Peperral (1947).[46] At the 10th meeting of Animal Husbandry Wing of ICAR, Zal R. Kothawala recommended crossbreeding not as a policy for universal adoption, but as a policy to be adopted in areas where the improvement of indigenous cattle through grading was not possible and in areas where there was a great scope for augmenting milk output. However, explaining how this policy got shifted suddenly in early 60s to crossbreeding cattle with European dairy breeds, they note:

In 1961, the committee set up by Central Council of Gosamvardhan recommended that crossbreeding of cattle be taken up *intensively using two or more exotic breeds in three selected areas, one each in plateau, plains and hills*. Similarly, Intensive Cattle Development Projects were started in 1964, as a part of special development programs in later half of the Third Five Year Plan. Crossbreeding of bovines through Artificial Insemination technique by introducing exotic breeds like Jeresy, Holstein Friesian, and Brown Swiss etc. to augment the milk production and to improve the breeding efficiency of bovine was given the prime importance under ICDP.[47]

The Working Group of Experts constituted to review the cattle breeding policy had also concurred with Central Council of Gosamvardhan recommendation of introducing crossbreeding in places where there are facilities for rearing and maintaining highly productive milch cattle, and around urban areas and townships to ensure adequate marketing facilities. Similarly, The Scientists' Panel on Animal Husbandry set up by the Ministry of Food and Agriculture in 1964 had also recommended crossbreeding in cattle on a priority basis in the intensive cattle development projects and in other milk shed areas.[48] Similarly, detailing how the shift in the policy on breeding came about, George and Nair (1990: 07) state:

The Board of Agriculture and Animal Husbandry wing of the Government of India (1958) and the Expert Committee appointed to evaluate Key Village Schemes (1959) recommended a shift of policy to crossbreeding of indigenous non-de-script cattle with exotic stock for rapid increase in milk production. Though, the Third Plan referred to the need to evolve better dual purpose breed for increasing the work capacity and milk production potential of the Indian cattle, its programmes gave high priority to crossbreeding for increased milk production.[49]

The strategy for selecting the area where ICDP was going to be implemented corresponded with the one adopted at the time of launching Intensive Agriculture Districts Programme (IADP), namely intensive area development approach initiated in June 1960 covering 15 agricultural districts containing over 20 million acres of cropland. George and Nair (1990: 03) remind us that each ICDP area covered roughly a lakh of cow and she-buffalo population, 'where good potential and conditions existed' for cattle development and milk production. The programme was expected to achieve a breeding coverage of about 70% of cattle population of each project and an increase in milk production by about 30% over a period of 5 years. In such areas, if KVS was under implementation, the components were merged into ICDP. In other areas, KVS continued in its original form.

At the beginning of Fourth Plan, 31 ICDPs and 490 Key Village Blocks were operating. During the Fourth Plan, the programme got expanded further and at the end of Fourth Five Year Plan there were 62 ICDPs in total. Corresponding with the green revolution phase in crop, this phase marks increased technologization of animal husbandry and dairying. George (1988: 03) reports that funds for pasteurization and associated technological improvements within the dairy sector accounted for about 43% of the total expenditure in animal husbandry and dairying! As a result, 22 liquid milk plants and 4 milk product factories in the organized sector had been operationalized during the second phase of ICDP and 25 milk supply schemes and 8 milk product plants were under various stages of completion. However, this increased thrust on processing technology and rapid integration with metropolitan milk markets meant that both crop cultivation and milk production had started to chart a journey towards separation and specialization. ICDP era initiated a process of gradual marginalization of concerns on draught power requirement in the farm sector and cattle development came to be interpreted largely as improvement of milk yields of cows and buffaloes and providing an impetus to push the milk supply and marketing to cities from rural hinterlands. It was assumed that mechanization would expand in a scale-neutral fashion and the bovine stock holding pattern would shift to increase in the proportion of milch animals at farm (operational holding) level. In the next section, changes in cropping pattern and bovine stock in study site are discussed in fair detail.

One of the ICDP sites was Kheda, where the implementation of ICDP was mediated through Kheda District Cooperative Milk Producers Union (KDCMPU), which had started to focus on the issues of veterinary assistance, artificial insemination services, animal nutrition and crossbreeding around this time.[50]

7.8 Green Revolution-Induced Change in Cropping Pattern

For the district of Kheda, the changes observed after the onset of green revolution show that progressive farmers very quickly adopted hybrid bajra. Table 7.1 shows areas under irrigated crops in Kheda district across a longer timeline.

Table 7.1 Irrigated crops in Kheda district (1971 till 2010)

Crop	1971–'72	1975–'76	1980–'81	1985–'86	2010–'11
Wheat	448	615	615	618	1545
Pearl millet	1603	1442	1603	1853	1616
Jowar	86	67	14	16	38
Rice	929	802	865	943	1011
Other cereals	373	456	NA	NA	NA
Total food grain	3439	3382	3099	3430	5550
Grams	14	13	30	28	
Tur	130	133	136	134	
Other pulses	97	104	85	113	
Total pulses	241	250	251	275	130
Cotton	748	773	473	538	350
Groundnut	214	191	188	190	
Tobacco	593	627	935	1133	1149
Total	1555	1591	1596	1596	

(in'00 ha) (for 2010–'11 the data also includes recently carved out Anand district) Source: District Wise Area, Production and Yield of Important Food and Non-Food Crops in Gujarat State, Several Years.

From Table 7.1, we can observe that the acreage under rice and wheat has been increasing—out of 220.90 thousand hectares of irrigated area, wheat and rice account for 80%—while coarse grains such as *jowar* are showing a declining trend. Another significant irrigated crop is cotton, which accounts for 11% of gross irrigated area. However, a crop that has remained a mainstay in Kheda district for quite obvious reason supporting the livestock-based economy of the region is *bajra* (pearl millet). So how does the district perform in terms of devoting the part of cultivated land for growing fodder crops? The latest statistics shows that the farmers here have kept devoting just 4 to 5% acreage to fodder crops. This could also be explained without surprise as from very early on KDCMPU have supplied cattle feeds (Table 7.2).

7.9 Green Revolution-Induced Changes in Bovine Economy

In 1968, an evaluation of five districts reveals how the stress on technological innovations to raise yields of cereal crops accompanied with advocacy for rapid mechanization led to interrelation between agriculture and livestock getting severed. In Ludhiana, for example, as a result of substitution of tractors for bullock power, installation of private tube wells accompanied by the extension of major irrigation and drainage facilities, the acreage under food grain crops went up by 14% in the year 1967–'68, yields of wheat and maize rising by 175% and output rising by 212%. Ludhiana farmers were now seen heavily investing into tractors after 1965. IADP studies indicated that one standard 35-horsepower tractor could easily replace 10 or

Table 7.2 Area under fodder crops and gross sown area in Anand and Kheda districts

Name	Area under fodder crops ('000 ha)						Gross sown area ('000 ha)					
	00–01	03–04	04–05	05–'06	06–'07	07–'08	00–01	03–04	04–05	05–06	06–07	07–08
Anand	10	16	15	15	15	15	195	294	298	307	297	306
Kheda	14	19	19	20	18	22	352	403	433	407	405	426

Source Dairying in Gujarat, A Statistical Profile 2013, NDDB, Anand, p. 97

more bullocks and facilitate the transfer of 45 acres of Kharif and 25 acres of rabbi land to other uses, if the bullocks and other maintenance cattle were removed. Similarly, Sarkar and Prahladachar (1966) in their study of 39 Dharwad district cultivators concluded that tractor owners reduced the number of bullocks maintained by 25%.[51] However, if we take the statistics for the year 1966, such a green revolution policy induced restructuring process in bovine economy doesn't appear to be a pan-Indian phenomenon. Even in Punjab and Haryana work animal population had been rising despite rapid spread of mechanization.

Quoting from a 1966 report of Automobile Association of Western India, Khurody (1974: 12) reports that there were 65,000 tractors in use and most of them were of 14 to 25 HP (horsepower) capacity. He also underlines the fact that as per Livestock Census 1966, there were some 38 million pair of working bullocks (34 million pairs of oxen and 3 million pairs of male buffaloes), with 39.8 million wooden ploughs, 3.5 million iron ploughs and 12.6 million farm carts. He adds that 1.9 million cows, generally sterile, were also used for work! Very strangely Khurody pulls out a counter intuitive observation from 1966 Livestock Census that "In spite of some half a million motor trucks plying for goods traffic, the number of animal driven carts actually increased from 12 million in 1961 to 12.7 million in 1966"! Similarly, 1966 census had also reported an increase in the number of bullocks operated sugarcane crushers from 5,90,000 to 6,50,000 in the same five year span. This clearly shows that an adoption of mechanization of agriculture implements in Western India had not displaced the role and functions of bullocks in agrarian processes as of mid-60s.

In a similar fashion, in each IADP district, it was not a similar tale of cultivators suddenly giving up the social attitudes around sacredness of cattle, in order to reduce the number of cattle by embracing cattle culling and slaughter programme with an improved fodder production and feeding programme. For example, an expert who studied IADP districts reports that a 1962 study for Raipur estimated that such "reduction in livestock numbers could have more than tripled milk production and released an estimated 700,000 acres of land for food grain production," in addition "to provide possibility for increased fecundity of cows and increased work capacity of remaining bullocks for agricultural operations". While following such expert advice would have reduced out of pocket expenditure on annual cost of replacing bullocks to a very low figure within five to ten years, Raipur cultivators neither adopted rapid mechanization nor did they find fodder cultivation an attractive option over cultivating paddy. Could such a differential place-based responses to expert advice be due to the already well developed tannery industry around Ludhiana district in 1960s, while the absence of it in and around Raipur? While in his study on IADP, Brown (1971: 70) recounts that "reproduction from cattle maintained in the district provided about half of the new bullocks needed every year" and thereby making cultivators dependent on importing another 80,000–90,000 bullocks annually for replacement, he remains oblivious to what could be termed 'mutuality' rather than one way 'dependency' by ignoring to reflect on the fate of traditional cattle breeders in neighbouring districts to Raipur who had been supplying Raipur cultivators the bullocks bred by them and who would go out of business (livelihood) if cultivators were to pay heed to expert advice.[52]

While the initial phase of dairy development policies laid greater bias for cows and neglected buffalo as milch animal, alongside the thinking over ICDP, there appears to be some attention given to recognizing the outstanding role played by the famed Murrah buffalo in milk production. In the year 1967, the government had decided to establish a new 200 ha stud centre at Nabha in Punjab for improving buffalo stock by crossing with Murrah. Whyte (1968: 214) had taken note of this scheme stating that the interesting popularity of the buffalo as the best milch animal in Punjab was pronounced in the livestock census (1966) numbers that reported 800,000 buffaloes, compared to 500,000 cows, whereas around a decade back the numbers of buffaloes and cows were almost equal for the state of Punjab. The rapid diffusion of green revolution technology appears to have been responsible in creating a change in outlook not only towards crop cultivation but also towards animal husbandry practices in Punjab. The replacement of bullocks by mechanization in Ludhiana (IADP district) and other regions in Punjab seems to have made it possible to invest into buying more buffaloes during 1960s. This also must have been accompanied by the emergence of milk markets in urban area and organized dairying. During the years between 1966 and 1977, milch animal population had grown by over 12% compared to less than 2% of work animals in Punjab.

Nath (1965: 722) points out that in his study village in Ludhiana, he observed a process by which cultivators were importing milch buffaloes, while keeping less number of uneconomic cattle. He was explaining such a shift in outlook through three step process whereby Punjab cultivators' shift to complete reliance on stall feeding and growing fodder crops was obtained. Nath states:

> First, with consolidation even the small areas of pastures which had been available earlier, were eliminated. Secondly, with the growing intensity of cropping, hardly any areas of fallow lands are available for grazing of animals. Thirdly, with increase in irrigation and in agricultural productivity, cultivators now grow much larger quantities of green fodder. Therefore, if the present phase of mechanization, i.e. electrification of wells is followed by the further step of introduction of small tractors, the presence of large number of surplus cattle will not prove an intractable problem. The introduction of tractors will be accompanied by progressive reduction in the number of cattle through the simple process of *cultivators selling their existing bullocks and not buying new ones. Buffaloes will continue to be kept however, for their milk and their numbers may even increase, if commercial dairy farming makes progress. The need for farmyard manure will then be met increasingly from buffaloes.*[53] (emphasis added)

Grewal and Rangi (1980: 120–125) surveyed the dairy development in Punjab during the period 1961 to 1978.[54] They claim that though these years marked rapid diffusion of green revolution technology resulting in increased yields in crops, the milk yields per animal did not register any significant change in the period. However, even in the absence of increase in milk yields, the total milk output during this period had registered an upward trend, from 2.1 million tonnes in 1970–'71 to 2.9 million tonnes in the year 1978–'79. The authors have shown that the buffalo population in the state had almost doubled during this period, while the number of local cows dropped sharply. A similar trend was observed in Haryana over the period 1956–1977.[55] It would have been better if the authors of these research studies presented the period by dividing that into two phases, namely, the pre-Operation Flood phase and Operation

Flood phase I years, so that the effect of green revolution technologies and the effect of Operation Flood could have been examined. However, what we clearly see from the results of NSSO report on Land and Livestock Holdings (1991–'92) is that the estimates of adult female bovine stock (used mainly for milk production and dairying purpose) owned per 100 rural households in Punjab and Haryana were much higher than the national average. This was so not due to higher number of milch cattle compared to national average (actually the number of cows was less), but because of very rapid rate of buffaloization. Look at Table 7.3 for the exact figures reported in NSSO report No. 407.

Similarly, green revolution crop choices and mechanization further accentuated the trend of preference for female buffalo as the milch animal in Kheda that we have recorded from the latter half of nineteenth century itself. Dorsten (1990: 66–67) recounts several research studies on agrarian situation in Kheda during the 1960s that talk about an increase in the absolute number and the relative share of milch buffaloes in the total bovine population, and a corresponding decrease in the number and share of bullocks.[56] Quoting the research study carried out by R. M. Patel (1982) at Agro-Economic Research Centre, Vallabh Vidhyanagar, who had compared the survey results from 19 villages of Borsad taluka in 1964–'65 with a follow up survey in 10 of those in the year 1969–'70; he states that a decrease in the number of bullocks and an increase in the average size of milch buffalo herd was observed alongside increased use of pump sets and tractors and an increase in the average milk yields of buffaloes. This increase in the absolute number and the relative share of milch buffaloes in the total bovine population in entire Gujarat state also kept growing very rapidly during the decade of 1960s, and in the latter half, this rise almost amounted to 3% points annually, as shown below (Tables 7.4 and 7.5):

Even as male cattle population registered a sluggish growth—and within this the rapid decline in breeding bull population suggested a rapid diffusion of Artificial Insemination technology—female cattle population decreased by 3% and female buffalo population registered a considerable growth at 7% within five years period, when intensification strategy was around the corner and KDCMPU's growth had attracted the national attention.

What this data tells us is that in Gujarat, the number of adult male cattle that were only being used for work was going down (castrated bulls), while those who were used for breeding only, or for breeding as well as for work was noticing an

Table 7.3 Bovine stock holding per 100 rural households in prime 'green' revolution states (1991–'92)

States	Cattle		Buffalo		Bovine	
	adult female	Total	adult female	Total	adult female	Total
Punjab	38	101	136	229	174	330
Haryana	33	83	124	203	157	286
India	45	145	33	59	78	204

Source NSSO Report No 407

Table 7.4 Number of cattle and buffalo in Gujarat (1961 and 1966)

Description	Livestock census (1961)	Livestock census (1966)	Percentage increase or decrease compared to 1961
Cattle (males over 3 years)			
Used for breeding only	14,757	10,488	– 28.9
Others	3,031,118	3,090,897	21.0
Total	3,045,875	3,101,385	1.82
Cattle (females over 3 years)			
Used for breeding only	1,822,683	1,761,424	– 3.4
Others	14,972	18,325	22.4
Total	1,837,655	1,779,749	– 3.15
Young stock	1,673,025	1,662,817	– 0.6
Total Cattle	6,556,555	6,543,951	– 0.2
Buffalo (males over 3 years)			
Used for breeding only	10,051	8243	– 18
Others	29,945	26,755	– 10.7
Total	39,996	34,998	– 12.49
Buffalo (females over 3 years)			
Used for breeding only	1,683,609	1,801,042	7.0
Others	18,768	18,811	0.2
Total	1,702,377	1,819,853	6.90
Young stock	1,174,208	1,285,581	9.5
Total Buffalo	2,916,581	3,140,432	7.7

Source Season and Crops Report for Gujarat for the year 1970–'71, reporting Livestock Census 1961 and 1966, p. 23

impressively increasing trend. Thus, what R. M. Patel had reported in his micro level study of 19 Borsad villages for the decade of 60s was continuing and represented the larger trend of mechanization not wiping out the role of draught cattle completely from agrarian processes in Gujarat. This also suggests that the indigenous breeds of cattle in Gujarat held a prime position, mechanization was being embraced by only well to do farmers with larger land holding and if the number of bullocks held by a farming household was changing hands, the cash income swiftly went into bringing in an adult female buffalo. Crossbred cow does not seem to have been the prime mover in the story of 'white' revolution in Gujarat in pre-Operation Flood years.[57] What should also surprise researchers on livestock issues is that adult female cattle (as well as adult female buffaloes) in 1972 census also came with a category titled

Table 7.5 Number of cattle in Gujarat, livestock census (1966 and 1972)

Description	Livestock census (1966)	Livestock census (1972)	Percentage increase or decrease compared to 1966
Cattle (males over 3 years)			
Used for breeding only	10,488	19,723	+ 88.05
Used for breeding and work both	14,805	26,603	+ 79.69
Used for work only	3,052,691	2,990,598	– 2.03
Others	23,401	19,649	– 16.03
Total	3,101,385	3,056,573	– 1.45
Cattle (females over 3 years)			
In-milk	812,957	897,493	+ 10.40
Dry	812,935	744,469	– 8.42
Not calved even once	135,532	159,130	+ 17.42
Working	10,026	13,508	+ 34.73
Others	8299	13,607	+ 63.95
Total	1,779,749	1,828,207	+ 2.42
Young stock	1,662,817	1,650,829	– 0.72
Total cattle	6,543,951	6,535,609	– 0.13

Source Report and Provision Figures of 11th Livestock Census for Gujarat State (1972), p. 11

'working', suggesting that intensification of cropping was taking place with arrival of irrigation projects!

If we look at the number of adult male buffaloes in Gujarat as per 1972 census, we notice an increasing trend there as well and within the categories those used for work only shows a very impressive 40% growth. What must be the reasons driving this? Could it be that the adult buffalo male was used to obtain carting services and connecting the village with nearby market through a non-motor mode of transport, like what we witness even today in Western Uttar Pradesh? Could it be the story of meat sector driving an ancillary profit deriving function from male buffalo stock? One aspect is certain that livestock owners in Gujarat were now turning the tract into a resource pool of famed buffalo breeds (the category of adult male buffalo used for breeding only showed a 5% growth, while the category for breeding and work both showed around 12% growth). May be the explanation is to be found again in the changing crop priorities under 'green' revolution policies. The area devoted for cultivation of paddy was growing and the male adult buffalo was found to be suitable for preparation of seed beds (Table 7.6).

This was a story repeating itself in major milk producing states and advanced milk sheds throughout North-Western India. In 1976, National Commission on Agriculture acknowledged this trend, when it reported that in states like Andhra Pradesh,

Table 7.6 Number of buffalo in Gujarat, livestock census (1966 and 1972)

Description	Livestock Census (1966)	Livestock Census (1972)	Percentage Increase or Decrease
Buffalo (males over 3 years)			
Used for breeding only	8243	8726	+ 5.86
Used for breeding and work both	2655	2971	+ 11.90
Used for work only	21,379	29,967	+ 40.17
Others	2721	4876	+ 79.19
Total	34,998	46,540	+ 32.98
Buffalo (females over 3 years)			
In-milk	1,015,937	1,089,566	+ 7.24
Dry	638,908	780,386	+ 22.14
Not calved even once	146,197	190,619	+ 30.38
Working	13,201	19,152	+ 45.08
Others	5610	4176	– 25.56
Total	1,819,853	2,083,899	+ 14.51
Young stock	1,285,581	1,479,905	+ 15.11
Total buffalo	3,140,432	3,610,344	+ 14.96

Source Provisional Figures of 11th Livestock Census for Gujarat State (1972), pp. 11–12

Gujarat, Haryana and Punjab; "the proportion of buffalo in the total bovine population is considerably higher than the all-India ratio".[58] The report further remarked that "Haryana and Punjab contain buffalo population exceeding that of cattle". This should have put to rest any doubts if the changes in herd size and livestock holding pattern had not started emerging as a result of 'green' revolution programmes, driving the wedge between farmers with larger landholding and those other landless, small and marginal farmers whose livestock relied more on commons and grazing, rather than on purchased or cultivated fodder.

The trend towards buffalo was even more entrenched in Kheda district, where the 11th Livestock Census (1972) figures showed the number of in-milk buffaloes to be ten times higher than the number of in-milk cattle! Compared to 1966 figures, the number of cattle had shown decline by 12.45% in 1972 census, whereas buffaloes had shown an increase by whopping 30.34%! Diffusion of crossbred cow—that much touted Operation Flood strategy—was still just catching up in the imagination of milk producers in Kheda and Tenth Livestock Census (1972) had not reported their figures separately. What three phases of Operation Flood were set out to unfold was another process wherein Gujarat will continuously report rising share of crossbred cattle, but till the late 1980s, the percentage of crossbred cattle in total cattle population in Gujarat had remained pegged at less than 1%. The next chapter looks at the numbers reported in livestock census for the period 1987–2012 for Kheda district to underscore that even as there was a ninefold increase for numbers of crossbred cattle, both in terms of population and production figures, it remains the old and entrenched story of buffaloization in the premier dairying tract of India, the Kheda district (Table 7.7).

Similarly, three other districts of Gujarat which had taken a lead in forming Anand pattern milk cooperatives were also reporting that even as the population of cattle had noticed a diminishing trend, buffalo stock amongst bovine was rising. Table 7.8 presents data for Sabarkantha, Mehsana and Surat districts from the livestock census for years 1972 and 1977 (Table 7.8).

Dorsten (1990: 74) describes what he calls a more general restructuring process in bovine economy in North-Western India, which was characterized by an increased emphasis on buffalo-based dairy farming. Dorsten identified both push and pull factors as playing a role in accentuating this process and listed on the production

Table 7.7 Number of cattle and buffalo in Kheda (1966 and 1972)

Description	Livestock census (1966)	Livestock census (1972)	Percentage increase or decrease compared to 1966
Cattle (males over 3 years)			
Used for breeding only	182	614	237.36
Used for breeding and work both	192,585	1619	
Used for work only		167,975	
Others		650	
Total male cattle	192,767	170,858	– 11.36
Cattle (females over 3 years)			
In-milk	37,841	17,167	
Dry		17,094	
Not calved even once		1811	
Working		421	
Others		189	
Total female cattle	38,133	36,682	– 3.80
Young stock cattle	51,849	40,002	– 22.84
Total cattle	282,749	247,542	– 12.45
Male buffaloes over 3 years	1449	2927	102
Female buffaloes over 3 years	246,091	298,482	21.28
Young stock buffalo	178,433	253,809	42.24
Total buffaloes	425,973	555,218	30.34
Total bovine	708,722	802,760	13.26

Source Report and Provision Figures of 11th Livestock Census for Gujarat State (1972), pp. 15–19 and Season and Crop Report for the Year 1970–71, Table VIII, reporting Livestock Census Data, 1966

Table 7.8 Growth of cattle and buffaloes in three districts of Gujarat (1972–77)

	Cattle			Buffaloes		
District	1977	1972	% Change	1977	1972	% Change
Sabarkantha	3,88,000	4,24,500	− 8.6	3,10,600	2,96,700	+ 4.7
Mehsana	2,91,300	2,98,300	− 2.2	4,79,900	4,76,600	+ 0.7
Surat	3,58,500	3,66,400	− 2.2	1,33,600	1,20,900	+ 9.7

Source Quarterly Bulletin of Economic and Statistics: Livestock Census (1977), Vol 20, 1980

Table 7.9 Availability of grazing land (1947–1997)

Year	CPR land[a]	Protected and unclassed land	Total[a]	Livestock population in millions	Area (Ha.)/ head of livestock	CPR/head of livestock
1947	70	30	100	280	0.36	0.25
1950–51	66.89	38.96	105.85	292	0.36	0.23
1960–61	48.82	35.27	84.09	336	0.25	0.15
1970–71	43.82	32.79	76.61	353	0.22	0.12
1980–81	42.23	35	77.23	416	0.18	0.10
1990–91	40.10	34.87	74.97	445	0.17	0.09
1997	38	34.80	72.80	467	0.15	0.08

Source Indian Agriculture in Brief and Indian Council of Forest Research and Education, 1995
[a] includes permanent pasture and grazing lands. Total is sum of protected and unclassed forests and CPR land

side, the disappearance of shared grazing lands, the continued fragmentation of landholding, the shift in cropping pattern from subsistence to cash crops and the introduction of mechanized draught power and simultaneous demand for milk in the nearby big cities as a factor on marketing and distribution side. However, Dorsten and other critics of Operation Flood had not underlined the proposal put forward by R. O. Whyte, who had envisioned a different sort of intensification. The next section would engage with that proposal and his dissenting voice in reference to the focus of Third Five Year Plan.

7.10 Whyte's Questioning of Third Plan Priorities

Whyte (1967: 18) in his book 'Milk Production in Developing Countries', had raised concerns about the misplaced stress upon a cereal-based nutrition policy that led developing countries to go for chemical fertilizer application with a view to raise the production of food grains from cereal crops. Whyte urged upon the animal production people and the human nutritionists to join the forces and "argue and prove the unfashionable thesis that the real criterion of intensification of production from cultivable land is the maximum production of human foods correctly balanced in terms of carbohydrates and proteins of plant and animal origin". He suggested that this will demand the evolution of a new farming systems, especially on the very best irrigated land, in which soil structure, organic matter content, nitrogen content and water holding capacity are provided by rotations with pasture and fodder crops and the full return of dung and urine to the cultivated land. It must then be argued that it is upon such soil that fertilizers will give their greatest and most economic effects in terms of increased yield of food crops for direct human consumption and of cash crops, thus making it possible to release land for the production of feeds and fodder.

At a time, when the agricultural science was paying an increasing attention to plant breeders working on cereal crops, Whyte (1961: 202) calls upon plant breeders to work on herbage and fodder plants and outlines the objective of such a plant breeder as follows:

> It is necessary to consider what the objectives of the breeder of herbage and fodder plants should be in India, in relation to present and possible future trends in grassland and fodder development. One must take account of the fact that ecological management will probably remain the most promising techniques to adopt in the improvement of natural grazing lands, and that seeds of the indigenous grasses may be needed in only a limited number of sites. One must also consider the great need to increase the acreage under cultivated fodder crops throughout the country, as part of a general trend towards more alternate husbandry or mixed farming, perhaps related primarily to dairy development projects.

Being an FAO expert, Robert Orr Whyte very much shared the developmental vision that prevailed in 1960s and argued for increasing the food and milk productions by embracing intensification strategies. The cultivable land was emerging as the site where expert scientific interventions must take place and alongside that experts in dairy development were envisioning a different sort of mix cropping. We call this a different model of intensification that marks a departure from what came to be the dominant mode of intensification leading developing countries in South Asia to 'mono cropping'. With an excessive focus on cereals, it created such level of production system transformations in certain regions that 'green revolution' came to be known as 'wheat revolution'!

Whyte (1961: 86) proposes three possible ways to fit fodder production for a dairy industry into the then prevailing farming system thus:

> The first is the introduction of fodder farming into tiny holdings that are characteristic of much of India; can we support this with economic or other arguments, or would it not be better for these little people to grow grain legumes to diversify their system of monoculture and thus produce some vegetable protein to balance their own diets?
>
> Second, there are the larger holdings, where it would be perfectly feasible in due course to educate these monoculturists to grow fodder in crop rotations, and to maintain animals of average productivity on each farm unit. These farmers would need to feel assured of the economic stability of the dairy industry.
>
> The third type of development is the dairy colony, whether the communitarian village type, or the larger municipal, government or military farm type. From the fodder point of view, there is no doubt that these can be made the most efficient and economic, although they may not lead to the improvement of farming system over a wide area. It is possible to apply all the modern techniques of fodder farming under the control of a specialist. If a village

council is considering the establishment of a concentrated dairy unit of this type on land formerly devoted to paddy or a similar crop, it should be easy to prove that the economic returns per acre is higher from the intensive dairy unit than from the paddies.

Whyte (1961: 07) recommends establishment of Central Fodder Production Board on the lines of Central Soil Conservation Board, under the auspices of the Ministry of Food and Agriculture and a similar institutions by all the state governments. Whyte (1961: 10) considers that by late 1950s, improvement of nutrition of domestic livestock had turned out to be the greatest problem that required urgent solution. Outlining the directions for the solution of this problem, Whyte (1961: 10) adds:

Prominent in any work on improvement of animal nutrition must be the development of the grazing and fodder resources, by managing and utilizing the grasslands correctly and by increasing the production of fodder from cultivated land. It has been fully realized that correct management of the natural grasslands, if it could be achieved, would be a great contribution to soil conservation; that the cultivation of fodder crops and more particularly of legumes in arable rotations and systems of mixed farming would improve the fertility and reduce the erosion of the soil, that the provision of adequate amounts of and a correct balance between proteins and carbohydrates would be the basis of an improved livestock industry and would provide more milk for better nutrition to a primarily vegetarian people, and better draught power for the cultivation of land.

Contrary to Whyte's wish list, the planners who wrote Third Five Year Plan had neglected to address concerns over grasslands and fodder resources.

Around the time when shift from KVS to ICDP happened, crossbreeding too was being promoted in a big way. However, when it came to the subject of 'improving natural grassland and promoting cultivated fodder crops,' even the Third Plan literature appears to have failed to appreciate the contribution that research and development work on this subject could potentially make. Whyte (1961: 01) had voiced this feeling on the first page of the second edition of his book 'Grassland and Fodder Resources of India' stating:

The literature on the Third Plan does not indicate a complete realization of the potential contribution which the grassland and fodder resources of India may make in the overall development of agricultural production. This literature refers to a planned increase in food production of 110 million tons by the end of the Third Plan period, through the provision of more manures and fertilizers, and of more irrigation facilities, and the implementation of the programmes of land reclamation, soil conservation and plant protection, multiple cropping and the use of better seeds, all leading to more intensive farming. But it appears that *the great potential contribution of grassland and fodder resources has been omitted from this catalogue, or has been taken for granted as an auxiliary item in the background, not deserving of a place in the front row.*[59] (emphasis added)

This was a serious charge levelled at planners, for the person who wrote these words was an FAO expert who had come to India as Grassland and Fodder Adviser to government. Whyte (1961: 135) was aware that ICAR had sanctioned a five year scheme in 1953 in order to study the regional ecology of grass cover and carry out a rapid reconnaissance survey of grassland in India. Two botanists, P. M. Dabadghao and B. D. Patil were sent to the USA and UK, respectively, to study modern principles and methods. The actual survey had begun in September 1954, and was continuing well till 1961, when Whyte voiced the above words. The concerns expressed by Whyte were also echoed by V. N. Amble in his paper, 'Feed requirement of bovines and possibilities of meeting them'. Amble (1965: 70–76) had showed the existence of around 59% deficit in Digestible Crude Proteins (DCP) and around 20% deficit in Total Digestible Nutrients (TDN). The same deficit conditions were projected for the next 15 years, in the event of continued focus of policy on increasing cereal-based inputs in dietary regime in the country.[60] Following strongly worded criticism on the neglect of fodder question in the Third Plan, in 1965 establishment appeared to undertake a course correction, when the Central Council of Gosamvardhan grudgingly constituted a committee to study the issue.[61] Some estimates resulted from the calculations of this committee surprisingly underscored how the intensification agenda of 'green' revolution was leading to competition between food grain (for humans) and residues (for animal nutrition). So while Burns had estimated availability of *kadabi* (straw or roughages) in 1945 to be 174 million tonnes, the committee report estimated it to be 130 million tonnes in 1965!

Compare this neglect on the fodder question by Third Plan with the enormous attention that it attracted at the eighth meeting of Animal Husbandry Wing of Board of Agriculture that met at Mysore in February 1949 (1951: 61), which passed the following recommendations by its Committee on Dairying:

> (i) As in the case of increased grain production special attention should be devoted by Government and experimental stations, to improving the quality of fodder crops both with regard to yield and nutritive value. (ii) Greater attention should be given to grasslands and forest grasses which during the flush season should be conserved in the form of silage hay and dry-grass on as wide a scale as possible. (iii) Ways and means must be evolved and the methods standardized to process dry fodders and straws available with cultivators, to make them more nutritious and palatable to cattle. (iv) In order to induce the cultivator to produce more fodder crops and hitherto for their cattle, they should be provided concessions with regard to land revenue, irrigation charges etc. (v) Whenever new land is brought under plough, or dry land put under irrigation, under any of the major projects, of land reclamation or irrigation in future, it should be made compulsory by legislation that an adequate area of land is devoted to fodder crop production for maintaining the required number of cattle, both milch and draught, in a satisfactory condition and (vi) It should also be made compulsory under the above item that the crop rotation adopted must contain a leguminous crop which should be available as fodder for cattle.

Similarly, the meeting had passed recommendations on concentrate feeds arguing to take "immediate steps to prohibit the export of oil seeds to provide edible cakes for cattle, as well as extraction of oil by chemical solvent and to discourage the use of edible oil cakes as manure". Whyte (1961: 199) talks about the work that was undertaken by three regional research centres of ICAR (at Anand, Bangalore and Haringhatta in Bengal) on the issue of animal nutrition. The research had begun at Anand, Bangalore and Haringhatta in February 1952, March 1954 and September 1954, respectively. Whyte indicates that reports were available from all three centres for 1958–'59 and earlier years, but had been marked as "restricted" and hence there was not much information about the research findings in public domain.

Quite contrary to Whyte's opinion, from 1960 onwards, we also witness calls for reducing the numbers of non-performing bullocks and also reduce the area of land under fodder crops.[62] The main difficulty in analysing the food versus fodder question during 'green' revolution years is that agricultural statistics from those years didn't present comprehensive data on production or availability of feed and fodder. The agricultural census reported only the area under fodder crops and pasturage. Whyte (1961) makes an early effort to present a survey of fodder and grassland resources in India, but neither does he explain the method of computation nor provides regional figures on availability and requirements of feed and fodder.[63] Whyte (1961: 35) actually states without mincing words that "the data on acreage of fodder crops are not reliable, as may be noted from the wide discrepancies which occur in these figures as between first and second edition". In the year 1961, when he was writing the third edition of his book, he stressed upon "an urgent need for fodder crops to be raised to the status of forecast crop, so that the present position and trends may be stated more accurately".[64]

There appeared somewhat better statistics on feed and fodder resources by the year 1976, when as part of National Commission on Agriculture (1976), a committee of experts had estimated total amounts of feed available from each crop, cultivated fodder and pasturage giving district level figures and recommended fourfold increase in land to be set aside for development of fodder and grassland resources.[65] A realization had also dawned upon animal nutrition researchers working at Indian Veterinary Research Institute that "for higher productions the schedules needed to be drawn up on the basis of nutrient requirement worked out by the National Research Council, USA (1971)"! In 1974, Ranjhan worked out general feeding schedules for better rearing of calves.[66]

Following intensification of agriculture in certain irrigated tracts in North India and thanks to mechanization that made farmers replace draught bullocks with milch animals, fodder crops emerged to occupy 30% of cropped area in parts of Haryana, Punjab, Rajasthan and Uttar Pradesh. Even as intensive cultivation and mechanization of agriculture under 'green' revolution policies changed the nature of crop cultivation and cattle rearing from pastoral mode of resource use to capitalist mode of resource capture and production, the grazing land continued to diminish. Here in Table 7.9, we get the figures for the period 1947–1997:

https://link.springer.com/book/9789819715596 However, the concerns on draught animal power system even in rain-fed areas continued to get a raw deal by policy makers and planners, which made Ramaswamy (1983) state that "their indifference is partly due to the mistaken impression that the use of draught animal power is only a passing phase in the development process; and that the mechanization of agriculture and allied operations would mean the virtual elimination of draught animals".[67] Ramaswamy (1985) estimated that it would take 30 million tractors to replace some 300 million draught animals on small farms.[68] Singh (1995) argues that in the regions that lie in the rain shadow of 'green' revolution, draught bullocks still remain a mainstay in small farming and estimates that there are some 70 million work animals still.[69] In Gujarat, while the number of male cattle is not showing an increase, while number of female buffaloes and crossbred cows are growing at faster rate, bullocks do remain still relevant as the number of adult male cattle does not show downward trend either.

Despite the enormous contribution of large ruminants to draught power, this aspect of crop-livestock system has not received much attention from researchers, except in late 1980s, when research papers highlighted the most critical issues facing the draught animals, probably under the influence of consistent advocacy by N. S. Ramaswamy, the then director of Indian Institute of Management, Bangalore. Besides Ramaswamy, there were two other brilliant researchers on draught animal energy, N. S. L. Srivastava and T. P. Ojha, both former directors of Central Institute for Agricultural Engineering (CIAE), Bhopal, and a national seminar on draught animal energy took place at CIAE, Bhopal in 1987.[70] In a paper that Ojha wrote for this 1987 seminar, he did not strike out the tractorization of farm sector, but held that tractors were "complementary source of power alongside draught power and will continue like that for many more years to meet the increased demand of power, of increased production and productivity".

The next chapter discusses the phase after that of 'green' revolution in crops and cattle, namely 'white' revolution or what has been called Operation Flood. The chapter in the earlier part discusses KDCMPU and other city-based milk schemes, in order to provide a backdrop on the dairy development agenda. The latter part would also review a selection from social science criticism of Operation Flood and suggested alternatives alongside reviewing the response from NDDB spokespersons to the criticism levelled on Operation Flood.

Notes

1. In 1942, Gandhi had established a pan-Indian organisation that would work on the question of cattle and farm economy, namely Go Seva Sangh at Wardha, which was later re-named as Go Krishi Seva Sangh.
2. In the early 60 s, the country reported a stock of 31 thousand tractors and by early 80 s, this increased to 663 thousand. The availability of tractors per lakh acres of gross cropped area increased from 20 to 385.
3. Sir Albert Howard, the imperial economic botanist had come to India in 1905 to work at Imperial Institute on Agricultural Research at Pusa on wheat breeding. He however, was open to appreciate and learn from the native wisdom and published what has now become a classic in the world of Organic Farming, An Agricultural Testament in the year 1940.
4. Dasgupta, Satishchandra (1945) *The Cow in India*, Vol I, Khadi Pratishthan, Calcutta.
5. Till 1926 there was little systematic work on artificial fertilizers in India and the relatively small amounts were used in case of plantation crops. In 1937 when E John Ruseel wrote the review of the work on Imperial Council of Agricultural Research, he noted that Southern parts of British India were more willing to adopt using chemical fertilisers compared to peasants from the North.
6. See, Bhatia, B M. (1970) *India's Food Problem and Policy since Independence*, Somaiya Publications, Bombay, p. 78.
7. Majumdar, N. A. (1960) 'Cow Dung as Manure,' *The Economic Weekly*, May 14, pp. 713–14.
8. Brown, Dorris (1971) *Agriculture Development in India's Districts*, Harvard University Press, Cambridge, Massachusetts.
9. See, Perkins, John H. (1997) *Geopolitics and the Green Revolution: Wheat, Genes and the Cold War*, Oxford University Press, New York, p. 158. Sir Henry Knight in his study, *Food Administration in India, 1939–1947* cites the cereal availability figures at 437 lbs in 1931–'32

and 382 lbs in 1939-'40; which appears to be higher. The reason being that Knight arrived at these numbers by taking food grain availability per adult member, rather than per capita.

10. For a detailed account of the policy level debates on the issue of Food Production and Food Security from 1942 to 1966 and a sharp critique of the minor role played by Bhakra Irrigation project, see Dharmadhikary, Shripad et al. (2005) *Unravelling Bhakra: Assessing the Temple of Resurgent India*, Manthan Adhyayan Kendra, Badawani, pp. 57–88.
11. In 1947, Lord Boyd Orr was invited to carry out a performance appraisal of Grow More Food Campaigns and suggested a new policy for India to attain self-sufficiency in food grains by March 1952 by making up for the deficit of 4.8 million tons.
12. See Agarwal, Shambhoo Nath (1943) *India's Food Problem and Price Control*, Business Bureau, Commerce Department, Allahabad University. Within less than two years of its coming into existence the Department of Food had held four all India conferences to discuss the food problem and policy. See, Proceedings of the Fourth All India Food Conference held at New Delhi on 13th, 14th, 15th and 16th October 1943, Government of India Press, Delhi, 1944 for the discussions in the wake of the report submitted by the Foodgrains Policy Committee. By January 1st, 1944 Department of Food had published A Manual on the Organisation of Food Control and Rationing. For a review of Food Administration prevailing during the period 1939–1947, see Knight, Henry (1954) *Food Administration in India, 1939–1947*, Stanford University Press, Stanford, California. The study was undertaken by Sir Henry Knight, a former ICS officer at the request of Food Research Institute at Stanford University and the Rockfeller Foundation had granted funds that made the preparation and publication of this study possible. Knight describes the period till the constitution of a separate Department of Food at New Delhi as a period where within the food administration bureaucracy the food problem in India appeared to be that of merely fixing prices.
13. Thanks to such a feeling the post-independent government laid great emphasis in imitating the China and sent the nascent irrigation bureaucracy to learn from the way this communist nation was implementing massive irrigation dams and embankment building projects.
14. Kolhatkar, V. Y. (1946) *Reconstruction of Indian Agriculture*, Popular Book Depot, Bombay
15. In the aftermath of The Bengal Famine, Gandhiji was frequently writing and speaking a lot on the question of food shortages. For a compilation of his views on the issue, see Gandhi, M. K. (1949) *Food Shortage and Agriculture*, Navjivan Publishing, Ahmedabad.
16. See, Perkins, John H. (1997) *Geopolitics and Green Revolution: Wheat, Genes and the Cold War*; Oxford University Press, New York, p. 168.
17. (1950) *Proceedings of the Seventh Meeting of Crops and Soils Wing of Board of Agriculture and Animal Husbandry in India* (held at Madras from the 7th to 10th of April 1948), Government of India Press, Shimla.
18. Jalan, B. L. (1951) *Food Problem in India*, Marwari Chamber of Commerce, Calcutta. Also see, Bansil, P. C. (1954) *India's Food Resources and Population*, Vora & Co Publishers, Bombay.
19. For a longer description of Community Development Programs, encouraged by development aid from US based philanthropist organizations like the Rockefeller Foundation and the Ford Foundation and then a shift from that more towards unfolding of Intensive Agriculture Districts Program (IADP) see, Rosen, George (1985) *Western Economists and Eastern Societies*, Oxford University Press, Bombay, p. 51, pp 71–80.
20. US Government and Ford Foundation provided more than $ 100 Million for Community Development Program, during the First and Second Plan period.
21. *Observations on Indian Agriculture*, A Report of United States Technical Cooperation Mission, New Delhi, March 16, 1953.
22. *Report of the Second Indo-American Team on Agricultural Education, Research and Extension*, Indian Council of Agricultural Research, July 1960. The report formally got published with a delay of five years.
23. For details on The Rockefeller Foundation's work in India's Agricultural development program, see Streeter, C. P. (1969) *A Partnership to Improve Food Production in India*, The Rockefeller Foundation Special Report, pp. 93–99.

24. However, in the very next year this bounced back to 73.5 million tons and in the last year of second plan output indeed reaching the revised target figure of 80 million tons. Actually, despite re-orientation to prioritise industries rather than Agriculture, the increase in food grain output in the last year of Second Plan over the last year of First Plan amounted to 14 million tons, while the increase during the first plan period was in the range of 10 million tons.
25. See Thorner, Daniel and Alice Thorner (1959) 'Ploughing the Plan Under: From Lathes to Ploughshares,' *The Economic Weekly* Republished in their, *Land and Labour in India* (1962).
26. While we concentrate here in analyzing policy discourses around the onset of intensification and 'Green' revolution, an analysis of the audio visual publicity material in the realm of agriculture and irrigation would be enlightening to understand the higher levels of technologism and crop husbandry being pushed in the direction of capitalist mode of production. A foreign company, Union Carbide that was producing pesticides and was responsible for India's worst environmental industrial disaster in Bhopal was seen claiming credit for "nation building" in early 1970s in print advertisements that it put in Economic and Political Weekly.
27. See *Five Years of Research on Dwarf Wheats*, Indian Agricultural Research Institute, New Delhi, INSDOC Press, 1968.
28. Report of the Co-operative Planning Committee, (appointed by Govt of India on the recommendations of the Fourteenth Registrars' Conference), 1946, Government Press, Bombay.
29. Annual Administrative Report of the Department of Animal Husbandry and Veterinary Science, Bombay, 1950–51.
30. For the details of a similar scheme that aimed at rehabilitation of Maldharis for the development of cattle industries in Saurashtra, Bombay state see, *Report of Community Projects and National Extension Service* (1957), Vol III, Part 1, Appendix 26, pp. 259–263.
31. Probably what he intended to refer to was the gradual erosion in the access of nomadic breeders of Gujarat to what earlier was available for grazing.
32. Whyte, Robert O. (1961) *Grasslands and Fodder Resources of India*, Indian Council of Agricultural Research, New Delhi.
33. Madan, G. R. (2007) *Cooperative Movement in India*, Mittal Publications, New Delhi, p. 241.
34. In his report on the Development of Cattle and Dairy Industry in India, N. C. Wright (1937) had voiced an opinion that Artificial Insemination technique had reached a stage in late 1930s, where it could be applied widely for upgrading the livestock in India. However, he was quick to point out that in India, "it is very questionable how far the method could be applied as a direct means of improving country stock: the difficulties of transport, the lack of communication, and the ignorance and probable prejudice of the cultivator would present almost insuperable obstacles". He had suggested that "the utility of the method ought to be investigated in some of the large Government cattle breeding centres, where adequate facilities are available and where breeding control is possible".
35. Chapter on 'Animal Husbandry,' *National Commission on Agriculture*, Brief Overall Report, (1976) p. 306.
36. Government of India (1961) *Review of Key Village Schemes*, Ministry of Agriculture, GoI, New Delhi.
37. Cited in Chakravarthy, A K (1985) 'Cattle Development Problems and Programs in India: A Regional Analysis,' *GeoJournal*, Vol 10, No 1, pp. 21–45.
38. Ensminger, M. E. (1956) *Research in Animal Husbandry*: A Review of Work Done during 1929–54, Indian Council of Agricultural Research, New Delhi.
39. See, *Interim Report of the National Commission on Agriculture on Milk Production through Small and Marginal Farmers and Agricultural Labourers*, Ministry of Agriculture, Government of India, December 1971, p. 37.
40. Such a suggestion was clearly at variance with the policy advocated by imperial dairy expert W. Smith and Norman C. Wright in his review report in 1937. Both of them had clearly uphold Denmark model of organising the processing and distribution of milk and milk products on co-operative lines. What prompted the experts to imagine cooperative organisation restricted only to handle milk procurement side, while bringing in bureaucratic government institution in dealing with the processing and marketing side remains a moot question.

41. Thus we can see that there had arose efforts to replicate Aarey model of milk colonies around cities – with or without acknowledgment of that model of rehabilitating the city milch stables by Dara N. Khurody at Bombay – sadly even with such intense social science debates during the second phase of Operation Flood, there is very little research work done on the fate of these other Aarey like milk colonies.
42. Das, K Madhava (1960) *E. M. Hough's Co-operative Movement in India*, Oxford University Press, London.
43. Writing about a village in Punjab and describing the technological change that it was undergoing during the early 1960s, V. Nath stated that the villagers had started to overcome the traditional prejudice against sale of milk. Nath states that cooperative milk producers' societies were organized in this part of Punjab, which used to supply milk to Government Dairy at the state capital. He also noticed amongst cultivators a tendency to increase the size of the herd by purchasing buffaloes from neighbouring districts. Nath also observes that the cultivators in this part of Punjab were not keeping 'surplus' or un-economic cattle at all. Nath remarks, ""there were 131 pairs of bullocks for over 1100 acres of intensively cultivated, well irrigated land, which has a high intensity of double cropping. *This number is not only not excessive; but appears to be barely sufficient for meeting the requirements of agriculture. The only other animals kept in appreciable numbers are milch buffaloes, which are the primary source of milk supply. On the other hand, the number of cows and young stock, which form such a large proportion of cattle in several other parts of the country, is comparatively small*". (emphasis added).
44. KDCMPU operated in a geography that had decades long history of being integrated into city milk supply to Bombay. During the initial years of operations, KDCMPU faced stiff competition from an established private firm, Polsons, in its efforts to win over the monopoly procurement rights to supply milk to Bombay Milk Scheme, which it won by the year 1952 only. In real terms, then KDCMPU's success story is more on the milk processing, distribution and marketing side and less on its 'unique' style of working on the milk production issues.
45. Chakravarthy, A K (1985) 'Cattle Development Problems and Programs in India: A Regional Analysis,' *GeoJournal*, Vol 10, No 1, pp. 21–45.
46. Peperral (1947), a British animal husbandry expert who studied the problem of 'dual purpose' breed, while agreeing with Olver, further maintained that under existing conditions in India, it would be difficult to maintain separate breeds of animals for milk and draught. He therefore stressed the importance of efforts to increase the milk yields of indigenous breeds of cattle rather than to improve the quality of the cows through large scale crossbreeding with imported bulls. The joint committee appointed by Indian Council of Agriculture Research and Sarva Seva Sangh in the year 1949 generally agreed with the suggestions of Peperral.
47. Rao, Venkatsubramaniam and Wit (1995) Consequences of crossbreeding of Cattle, *Economic and Political Weekly*.
48. See, *Interim Report of the National Commission on Agriculture on Milk Production through Small and Marginal Farmers and Agricultural Labourers*, Ministry of Agriculture, Government of India, December 1971, p. 38.
49. As we have noted earlier, the priorities and programme designs were changing by the latter half of Third Plan period both in crop as well as animal husbandry programmes.
50. See, Patel, Sujata (1990) 'The Anand Pattern: Its Origin and Growth,' in Doornbos, Martin and K. N. Nair (1990) *Resources, Institutions and Strategies: Operation Flood and Indian Dairying*, Sage Publications, New Delhi.
51. For details account of policy level arguments on the mechanization of agricultural operation in Punjab, see Kahlon, A. S. and D. S. Gill (March 1967) *A Case for Mechanising Selected Agricultural Operations in Punjab*, Directorate of Economics and Statistics, Ministry of Food and Agriculture, Government of India.
52. Other researchers have argued that despite the planners' and development official's advice to the contrary, farmers' local practices at times have been able to find the value in retaining crop – livestock interrelations and synergies, thanks to their ability to combine development inputs across different programmes. For example, see Basu, Pratyusha and Bruce A. Scholten

(2012) 'Crop-Livestock System in Rural Development: Linking India's Green and White Revolutions, *International Journal of Agricultural Sustainability*, Vol 10, No 2, pp. 175–191.

53. Nath, V. (1965) 'The New Village II: Technological Change in Agriculture', *Economic and Political Weekly*, April 14, 1965. Pp. 713–722.
54. Grewal, S. S. and P. S. Rangi (1980) 'Economics and Employment of Dairying in the Punjab,' *Indian Journal of Agricultural Economics*, Vol 30, No 4, (October-December), pp. 120–125.
55. Katariya, M. S. (1982) *Geography of Indian Livestock and White Revolution*, Vishal Publications, Kurukshetra.
56. Dorsten, F. Van (1990) 'The Impact of AMUL on the Milk Economy of the Kheda District,' in Doornbos, Martin and K. N. Nair (eds) *Resources, Institutions and Strategies: Operation Flood and Indian Dairying*, Sage Publications, New Delhi.
57. Actually, in the next chapter it has been argued with evidence coming from NSSO data and Livestock census data that crossbred cow as a mirror image of HYV seed story functions more in 'discursive' domain, while in the Anand pattern, it has been a story of continued buffaloisation. The diffusion of crossbred cow has been more in states, where agricultural operations doesn't put value on bullocks and there is only one such state in South India, namely Kerala where thanks to indiscriminate crossbreeding, the crossbred stock in cattle amounts to more than 90 percent as per latest livestock census figures.
58. See, *Report of the National Commission on Agriculture*, Part VII, Animal Husbandry, Ministry of Agriculture and Irrigation, New Delhi, p. 66.
59. Although in the same chapter, Whyte (1961: 07) had noted elsewhere that the research work on the subject of natural grassland and cultivated fodder crops was receiving attention for last 10 years and under the Third Five Year Plan, Planning Commission had just taken a decision to approve the proposal submitted by the Ministry of Food and Agriculture for the establishment of an Indian Grassland and Fodder Research Institute and its five sub-stations.
60. Amble, V. N. (1965) 'Feed requirement of bovines and possibilities of meeting them,' *Indian Journal of Agricultural Economics*, Vol 20, No 1, (Jan-March), pp. 7–76. Also see, Amble, V. N.; Murthy, V. V.; Sathe, K. V.; and goel, B. B. P. S. (1965) 'Milk production of bovines in India and feed availability,' *Indian Journal of Veterinary Science and Animal Husbandry*, No 35, pp. 221–33.
61. See *Report of the Committee on Livestock Feeds and Fodder*, 1965, Central Council of Gosamvardhan, New Delhi.
62. See, Shastri, C. P. (1960) 'Bullock Labour Utilisation in Agriculture,' *The Economic Weekly*, October 29, pp. 1585–92.
63. Whyte, R. O. (1961) *The Grassland and Fodder Resources in India*, Indian Council of Agricultural Research, New Delhi.
64. Did the establishment pay any heed to this advice by National Grassland and Fodder Advisor? Fifteen years after Whyte wrote those angry words, here is what National Commission on Agriculture reported in 1976: "No accurate estimates are available on the amount of green forages cultivated annually. The joint ICMR and ICAR Committee indicated a supply of 111 million tons in 1954 for all livestock. The Central Council of Gosamvardhan estimated the availability of cultivated fodder at 132 million tons in 1961. Whyte and Mathur gave in 1965, an estimate of 142 million tons of green fodder availability for all bovines and changed it to 307 million tons based on IARS sample surveys. According to available information the area under fodder crops in 1970-'71 was 6.18 million hectares, out of which 1.02 million hectares was irrigated and cropped twice. No record is available of the type of fodder species grown. In the absence of data on acreage under different crops it is difficult to get any realistic idea about the total yield of fodder as well as of the nutrients provided to the livestock from cultivated fodder"!!!
65. For a spatial analysis on fodder availability based on these estimates read alongside 1971-'72 Livestock census figures see, Chakravarti, A. K. (1987) 'Availability of Cattle Fodder in India,' *Geographical Review*, Vol 77, No 2, pp. 209–217.
66. See, Ranjhan, S. K. (1974) *Feeding of Cattle, Sheep and Goats*, Animal Nutrition Division, IVRI, Izzatnagar.

67. Ramaswamy, N S (1983) 'Draught Animal Power,' *India International Centre Quarterly*, Vol 10, No 3, pp. 309–320. N S Ramaswami was an ardent advocate for supporting the draught animal power system with innovations and earned the epithet Cart man of India for his work.
68. Ramaswamy, N. S. (1985) *Draught Animal Power: Socioeconomic Factors*, in Copland, J. W. (ed) *Draught Animal Power for Production*, ACIAR Proceedings, ACIAR, Canbera, Australia, pp. 20–25.
69. Singh, K (1995) 'Research Priorities in Improving Animal Agriculture by Agro-Ecological Zones in India,' in Devendra C., Gardiner, P. (Eds.) *Global Agenda for Livestock Research: Consultation for South Asia Regions*, ICRISAT Asia Centre, Patancheru, 6–8 June 1995, ILRI, Nairobi, Kenya, pp. 85–101.
70. '65 Million Draught Animals,' *Down To Earth*, (October 31) http://www.downtoearth.org.in/node/11947

Chapter 8
Replicating Anand Pattern Milk Cooperatives: Studying the Operation Flood

Replicating a successful dairy cooperative model to enhance the milk production in India seems an innocuous proposition. Around the same time as the cattle development policy was witnessing a shift to *intensification*, the then Prime Minister Lal Bahadur Shastri while inaugurating a cattle feed plant established by Kheda District Cooperative Milk Producers' Union (KDCMPU) called upon a dairy technocrat to the task of replicating KDCMPU success story across the nation. Was Shastri the first one to speak about the need to replicate the success of the cooperative milk union model across the nation? Could we find such an emphasis in the policy discourses on dairy development during late colonial India? Was that statement by the then Indian Prime Minister a result of successful lobbying and advocacy efforts by young dairy technocrats working at KDCMPU and envisioning a pan-Indian milk and milk products marketing institution? This chapter brings years that witnessed white revolution under study along with the focus on these questions.

KDCMPU got its first milk processing facility in the form of the part of Government Research Creamery building and this was followed by inauguration of plate pasteurizer on 14 February 1950 with the daily processing capacity of 57,000 L. Then in 1955, another plant was added, and the cooperative invited Prime Minister Nehru to Anand to inaugurate it. This tendency to invite President or PM and then to attract a considerable media attention had helped the dairy technologists and managers at KDCMPU in their aim to drive the 'replicate the Anand pattern' advocacy a great deal.[1] Two days before Jawaharlal Nehru inaugurated the dairy plant owned by KDCMPU at Anand, *The Economic Weekly* in a very short editorial had compared Anand and Aarey models, by stating:

> That the Anand dairy should have been set up under the ownership of a cooperative union is a tribute to the leadership of the locality and *at the same time an example to milk producers of the other localities. It opens a new perspective for the solution of problems of rural areas in so far as they can be linked to the production of milk or eggs or other subsidiary foods.* Not merely the marketing of these products, but also their processing can be carried out along cooperative lines as has been done so successfully in Denmark. *If the example of Kaira and the Anand dairy is followed in other areas and other spheres, we ought to see the rapid*

H. Upadhyaya, *Critical Insights on Colonial Modes of Seeing Cattle in India (1850–1980)*, Asia in Transition 27, https://doi.org/10.1007/978-981-97-1560-2_8

> *development of a cooperative sector in the country, a force then would act as a balance to the expanding power of the State in the economic sphere and which at the same time would not be the private sector in the strict sense.*[2] (emphasis added)

Within six months after that, this weekly carried another article on KDCMPU success story. This article narrated the story of KDCMPU's growth, stating that the administrative affairs of the union are dealt with by an elected body consisting of representatives from village societies. This article informed readers that this elected body is advised on technical points by *two experts who had training in dairying in United Stated and New Zealand,* a reference to Varghese Kurien and H. M. Dallaya respectively, both of whom by then had not grown so big as to deserve mention by spelling out their names or probably deliberately avoided being shown in media limelight by conceding that space to milk producers and Tribhuvandas Patel![3]

However, Varghese Kurien was actually planning a big-ticket intervention in India on the front of marketing of milk products. In his first major speech at Anand in 1957, Kurien had talked about marketing of milk products in India. This speech is said to have provided the blueprint of Indian Dairy Products Marketing Board which Kurien conceived circa 1955. This speech, which was later published as a booklet by KDCMPU, articulates the thinking of Kurien (1957: 03) who argued that "the best way in which one can increase milk production is by arranging to market in an organized manner whatever milk is at present produced, either as liquid milk or as milk products". Thus, he held a viewpoint that looked at processing and marketing intervention as the vehicle for dairy development, as this had a potential to put in milk producer's hands higher returns on her product and would potentially encourage farmers to pay more attention to their milk production. In March 1957, KDCMPU had published his outline of a plan titled, *Marketing of Milk and Milk Products in India* wherein he made a fervent argument for establishing "Indian Dairy Products Marketing Board without much delay".[4] He advocated that such a dairy products marketing board be accorded the sole right of importing dairy products in India and wished that "one of the first things to be investigated is therefore the possibility of obtaining such a monopoly". Kurien had listed out eight main functions of such a marketing board as follows:

1. To take over the entire production of all the approved dairy products factories in India.
2. To pay for them on the basis of their quality as determined grading and to support the Agmark scheme.
3. To arrange for the storage, transport, distribution and marketing of these dairy products through common brand names.
4. To channelize through this board the import of foreign dairy products into India.
5. To pool the price of the various types of dairy products and to fix the purchase and selling price of dairy products.
6. To regulate the production of different types of dairy products according to the market needs.
7. To enforce quality standards.
8. To carry out propaganda for increased consumption of milk and milk products.

In addition to the above main functions, such a board would also take upon itself to perform certain additional functions, such as:

1. To finance research.
2. To assist in planning and execution of new dairy projects.
3. To arrange to import dairy machinery and to maintain servicing and spare parts facilities to all its component dairy factories.
4. To initiate the progressive manufacture of dairy machinery in India.
5. To maintain close liaison between the industry and the National Dairy Research Institute.
6. To generally protect the legitimate interest of the dairy industry.
7. To arrange for technical training in the field of dairying.

Contemporary sources from this period of 1950s had repeatedly compared economic performance of 'Anand pattern' with that of 'Aarey experiment,' while paying scant attention to a fact that Aarey had emerged from the stated government policy to rehabilitate city stables in peri-urban areas to improve city milk supply. V. A. Philipose of the Town Planning Organization in the Ministry of Health wrote in 1957:

> Aarey is an example of the nationalised sector of industry whereas Anand is an example of the cooperative sector of industry. Aarey stands for the limited purpose of the city milk needs, but it depends on the village cooperatives for a considerable portion of the needed quantity. Anand stands for the purpose of both milk supply and comprehensive development of village economy with the prospects of producing the entire quantity of milk for the city on their own, if needed and at the same time providing excellent opportunities for subsidiary occupations for the farmers. The organisation, centering in Anand is ideal insofar as the relationship of farmers and management is concerned, and it is set in an ideal grove promising really bright future and possibilities of all round economic advancement for the people. The above comparison raises in anyone's mind the question whether India's general economic position would permit such a huge financial investment for a project like Aarey with its extent of utility, when, with a very small investment, such splendid results as in Anand, could be achieved.[5]

Whether Philipose followed this contention by presenting side by side the investment made by State is not known. However, notwithstanding his claims of unimpressive milk production figures of Aarey, a paper that Khurody (1973: 14) himself wrote in 1973 reminds us that "at Aarey, the total production of milk from some 16,000 animals was 1,20,000 L per day in winter, or some 7.5 L per animal, per day". Khurody asserts that such a milk production would probably denote the highest recorded yield for the largest number of animals, at any place in India and reminds his readers that it was also the largest herd in the country, where milk recording was possible and was done. Khurody also claimed that "many buffaloes at Aarey gave an average of 12 L per day and the fat content was never less than 7%". However, writing fourteen years after that uncharitable comparison by Philipose, Khurody was humble to also admit that Aarey producers were not able to keep up that performance and states:

> Unfortunately, the efficiency of such a prized herd, through the neglect and negligence on the part of the authorities, has been brought down drastically during the last 7 or 8 years. The

> average is now (i.e. in 1973) only some 4.5 L per animal per day and there are only 10,000 animals or so. The total production has dropped to 45,000 L. The rest have left because of the unbusinesslike (sic) policy of the authorities. No one is prepared to come to the colony out of some 100,000 buffaloes which are housed in suburbs under the most difficult conditions for the cattle owners and attendants.

Having recounted such an experience of Aarey milk colony, Khurody (1973: 15) suggested that one possible approach out of such an impasse could be to "combine Aarey, Haringhatta and Madhavram and put them under a central autonomous authority to operate them efficiently".

Even during late 1980s, when Operation Flood had come under rigorous academic scrutiny, we witness Sujata Patel (1990: 41) trying to question what to Philipose was 'small investment' in case of Anand by relying on the figures of financial assistance to KDCMPU for the period 1949–1956, as reported in Kelly and Singh (1981: Table 3.2 and 3.9). Patel contends whether an annual financial assistance amounting to Rs. 3,00,000 during the period 1949–1956 could be termed as 'small investment'. Dara N. Khurody, who established Aarey milk colony, quotes the figure of Government of Bombay's financial assistance for dairy development in Kheda district between 1949–'50 and 1956–'57 at US $ 318,157. Khurody also stated that in addition to this, the Government of Bombay spent a sum of Rs. 2 million (or US $266,000) for infrastructure construction, viz. improving existing roads and building a few new ones.[6] Another flaw in Philipose's argument that missed even Patel's attention was his effort to compare apples with oranges. Aarey housed those cattle owners who had been displaced from the city stables and they were not peasants who had access to cultivable land. To evaluate Aarey, without acknowledging the role, it played in rehabilitating the owners of city cattle stables would be to miss the point that Aarey had emerged after a series of intervention by the Bombay government to move city cattle stables outside the municipal limits.[7] Popular narratives on KDCMPU paint Khurody and Aarey often in poor light, so much so that one fails to understand how the trio—Tribhuvandas K. Patel, Varghese Kurien and Dara N. Khurody—would have felt when the prestigious Ramon Magsaysay award was conferred on them jointly.[8] What becomes clear from these narratives cited above is that Kurien and Khurody did not always think alike and Khurody's model of milk colonies on peripheries of cities as a scheme to rehabilitate cattle owners from city stables did not find much support from Kurien[9]. Khurody's training was around the Military Dairy Farm model—his father worked on a military grass farm at Mhow in present-day Madhya Pradesh—and he was preoccupied with the question of how to rehabilitate the cattle and buffalo housed in city milch stables. He planned Aarey milk colony in a fashion whereby the milk colony would try to close the nutrition cycle: the manure collected would go as input to the grass and fodder farm. Milch cattle and buffaloes on the farm would be supplied with the fodder grown there. Contrary to this, Kurien was working in an ecosystem that had practised mixed farming and had shown buffalo-based dairying thriving thanks to market linkages with Bombay.

In 1960, when KDCMPU got its new dairy plant to manufacture cheese and baby food, an editorial in *The Economic Weekly* had stated, "Unfortunately Anand is not being replicated in every state and every district as it should be"[10]. As it becomes obvious, this idea that Anand pattern milk cooperative shall be replicated keeps resurfacing throughout the decade of 1955–1964. Thus, linear narratives which present the genesis of National Dairy Development Board (NDDB) as the suggestion that followed what the then Prime Minister Shastri witnessed in a milk cooperative village on the eve of his public speech at Anand have failed to engage with the sustained advocacy and lobbying efforts by KDCMPU, Varghese Kurien, H. M. Dallaya and his friends.

However, as it has been discussed in earlier chapters in this book, an intellectually exciting question to probe would be to recall the role played by early history of milk producers' cooperative unions and to enquire if there were calls for replicating the successful milk cooperative model prior to that speech by Lal Bahadur Shastri, calling upon the young dairy technocrat to replicate Anand Pattern Milk Cooperatives in 1964. To take a recap, in January 1926, speaking at the agriculture section of Indian Science Congress, Imperial Dairy Advisor, W. Smith had made a plea for cooperative dairying in India. He cited in the paper he presented, Calcutta Milk Cooperatives Union as a very successful example. He argued that "*there is no reason whatever why the example of Calcutta Cooperative Milk Societies Union should not be repeated all over the country*" (emphasis added). The discussion that had followed Smith's presentation—although not reported verbatim but coming mediated to us through the rapporteur's notes—is very interesting since there was a great amount of curiosity to understand what constituted the 'Calcutta model'. Responding to some questions, Dr. Harold H. Mann, who was chairing that session, stated that the primary difference was that near Calcutta, the small cultivators produced fodder for cattle and did not go in for cultivation of crops antagonistic to cattle keeping. Thus, Mann had suggested that the villages around Calcutta had integrated cattle keeping and milk production function in their cropping patterns. Rapporteur had also noted that during the discussion, president stated, in somewhat lamenting tone, that the discussion had been practically confined to city milk supply, and he indicated the possibility of cooperative milk production in purely rural area. The president also spoke about "the difficulty of dealing with "flushes" of milk at the commencement of the rainy season and the *problem of itinerant herds of dairy cattle*" (emphasis added).

In the year 1927, W. Smith had penned a brief note on the organization of Cattle Improvement Societies which Gandhiji later published in his magazine *Young India*. A decade later, in 1937, N. C. Wright in his report once again stressed the need to explore the possibility of cooperative milk production in rural area for the rural consumption. Wright had cautioned against trying to replicate the Military Dairy Farm type of milk production and factory system of processing. In his detailed review of milk and indigenous milk products, he suggested that there was a scope for increasing the consumption of fresh fluid milk. Although Wright had suggested that Anand research creamery could be closed down, his reasons for this suggestion have not been argued out in details. The only probable argument that we find in his report is that ICAR was already planning a dairy research institute in North India and hence he wished that research creamery should be located close to that institute. However, that suggestion was challenged by Imperial Animal Husbandry Expert, Sir Arthur Olver who argued for implementing a scheme to develop this research creamery at

Anand further. Olver also made a fervent plea to sanction a grant of Rs. one lakh and wished that the research creamery that was lying idle at Anand shall be reopened and a competent person shall be appointed. Olver was so much invested in resisting efforts to close research creamery at Anand that he proposed that the person to be appointed there shall be made head of Dairy Technology section at the new dairy research institute of ICAR, when that institute takes shape in North India.

Popular narratives on the dairy development of India that have relied extensively on interviews with Dr Varghese Kurien present repeated closing down and reopening of this government research creamery at Anand rendering it as some sort of joke. However, an honest engagement with archives and contemporary sources unveils the special place that the government research creamery at Anand occupied in the vision of imperial dairy expert, W. Smith and imperial animal husbandry expert, Sir Arthur Olver respectively. Both have gone on record with their opposition to proposed termination of government research creamery at Anand at different times.

8.1 The Era of City Milk Schemes (1950s and '60s)

In the seventh year of its establishment (i.e. 1956), Aarey milk colony was producing 43,000 gallons a day, a supply sufficient for half the population of Bombay then estimated at about 1,500,000. In its seven-year journey, Aarey was assisted by the Colombo plan countries, Food and Agriculture Organization (FAO henceforth) and other organizations. In the year 1956, there was a proposal for USD 1,000,000 expansion of Bombay Milk Scheme. FAO had sent four technical assistance experts to Aarey in addition to awarding fellowships to six of the milk colony staff members to get training in dairy science abroad. FAO had also organized a training course on tropical dairy farming, which was attended by 44 participants from 11 countries[11]. However, in the late 1950s, there were also some critical voices that articulated discontent on India getting heavily dependent on FAO assistance for dairy development. For instance, in the views of an article in *The Economic Weekly*, "in the year 1956, India, agreed to offer itself as a 'guinea pig' for the experiment on how to utilize agricultural surpluses accumulated in developed countries for helping continuous development in underdeveloped countries subject to the condition that disposal of surplus do not depress either domestic or international prices"[12]. FAO had published results of a pilot study in India in June 1956 and the first programme to use surplus agricultural products under the auspices of FAO was expected to be in action before the end of the year, specified a press release issued in September 1956. The product concerned was primarily dried skim milk and the supplying countries were the USA, Australia, New Zealand and Holland. By then the first scheme that had gone through was Calcutta Milk Scheme and similar schemes for assistance to other dairy centres in India were under contemplation. Beside dried skim milk to be supplied, there was a grant of 175,000 lbs or its equivalent towards equipment and expert advice for using milk powder to make toned milk. At that moment in September 1956, Calcutta had its milk colony at Haringhatta, which was similar to

Aarey and another smaller colony had been started in Fulia in addition to a proposal to establish a modern dairy farm at Kalyani Agricultural University. How this scheme of donated surpluses would change the milk production throughputs in Calcutta can be gauged when we compare that Haringhatta plant was able to use only 120 tonnes of dried skim milk per year, as against 1000 tonnes that Aarey milk colony was able to use in 1956. The dairy factory to be set up at Calcutta under the auspices of FAO was supposed to initially consume 1900 tonnes of dried skim milk and later its capacity might be extended by another 1000 tonnes[13].

In 1959, the Indian Government set up Delhi Milk Scheme on the lines of Aarey at Bombay and Haringhatta at Calcutta. During the late 1950s, the question of city milk supply was studied through the milk marketing surveys carried out by Dara N. Khurody. Multilateral assistance was now available for the construction of milk processing plants in the country that had gained independence just about a decade back. Reviewing the livestock development policy followed by India during the first two five year plans (1951–1961), following were the thrust areas and projections:

- To increase milk production.
- To improve milk supply to urban areas.
- City cattle stables to be brought under per-urban cooperatives supplying milk to Government Dairies.
- To improve quality and supply of draft animals.
- Grading up of indigenous breeds to improve the milk supply.
- Fodder scarcity to be tackled.
- Key village scheme to be launched.
- Programme of castration of scrub bulls to be taken up.
- Artificial insemination programme to be taken up and breeding of livestock to be regulated.
- Diseases of animals to be controlled.

Contemporary sources list increased milk production and improved supply to urban areas as well as investments made in establishing veterinary hospitals and dispensaries as achievements of the first two plans. However, the first two plans had failed to address the issues around improving the quality and supply of draught animals. The problem of large number of cattle in city stables continued and fodder scarcity had remained a problem still. As this study pointed out in previous chapter, even the Third Five Year Plan had failed in addressing fodder scarcity, improving supply for urban milk schemes and implementing policy on draught animals.

8.2 Milk Production Targets in the Late 1960s

While formulating the cattle development programmes for the Fourth Plan, the Union Ministry of Agriculture and Irrigation expressed its concern about the lack of necessary resources with most of the existing livestock farms looked after by respective state governments, for attaining progressive genetic improvement methods. The

Ministry, therefore, decided to set up six large cattle breeding farms on its own. Accordingly, the establishment of four cattle breeding farms, one each for Sindhi, Tharparkar, Jersey and Holstein breeds and two buffalo breeding farms, one each for Murrah and Surati, was taken up[14].

Whyte and Mathur (1968) describe the whole situation as it existed in the late 1960s on the question of how much milk does India need to produce if she wants to attain the objective of providing 280 g or 10 oz of milk per capita per day, a standard suggested by the Nutrition Advisory Committee of the Indian Council of Medical Research. That figure was stated to be 133 million litres milk per day. Since this was considered a very ambitious target, they had also worked out a more realistic figure of increasing India's milk production to the level of 80 million litres per day, if arrived at by assuming 170 g or 6 oz of milk per capita per day[15]. Datar Singh, then chairman of Animal Husbandry Wing of Board of Agriculture in 1949 had reported that "the average quantity of milk available per head of human population was 5.5 oz per day in undivided India, and corresponding figures for India and Pakistan respectively in post-partition situation were 5.2 oz and 6.8 oz". Thus, partition had effectively left India devoid of some of its best dairy breeds of cattle which resulted in a drop in milk production from an average of 646 lbs to 487 lbs per cow and from 1229 lbs to 1138 lbs per buffalo[16]. Datar Singh had wished that India shall be in position to increase per capita milk availability to at least half the minimum nutritional requirement, viz. 8 oz However, we need to compare these suggestions by Datar Singh in 1949 and by Whyte and Mathur in 1968 with the per capita milk availability circa 1937 as reported by N. C. Wright in his 'Report on the Development of Cattle and Dairy Industries of India', viz. 7 to 8 oz Thus, it appears that milk production level had simply not kept pace with the population increase in the decades of 1940s, 50s and 60s![17]

Chatterjee (1960: 1347), citing data from 'Indian Livestock Statistics 1952–'53', stated that per capita milk availability per day stood at 5.26 Oz Chatterjee also gave a breakup in terms of state level figures prior to reorganization of states and found that two states, namely Kerala (1.27 Oz) and Assam (1.9 Oz), showed per capita milk availability figures that were less than one-fourth of all India average. West Bengal (2.43 Oz) and Orissa (2.38 Oz) had reported a little less than half of all India average. Chatterjee showed that a few states reported per capita milk availability that was well above the ICMR-prescribed 10 Oz per day level. These were: Himachal Pradesh (20.4 Oz.), Punjab (17.43 Oz), PEPSU and Saurashtra (both at 14.4 Oz), Rajasthan (12.31 Oz) and Ajmer (10.2 Oz). At regional level thus, the per capita milk availability showed a remarkable variance[18].

As per the Fourth Plan estimates, the milk production in India had increased marginally from 20.0 million tonnes in 1966–'67 to 21.20 million tonnes in 1968–'69. The plan estimated that milk production would reach a level of 25.86 million tonnes by the end of the Fourth Plan period. The draft Fifth Plan had estimated milk production at 23.20 million tonnes in the year 1973–'74, which translated into per capita milk availability at 110 g in 1973–74. Around the same time, the 30th round of NSS estimated milk production to be around 20 million tonnes. The minimum nutrition requirement recommended in contemporary literature had put the figure at

201 g per capita[19]. Thus, clearly the Planning Commission estimation about the milk production had not assumed the rapid growth that Whyte and Mathur (1968) were projecting during late 1960s. It is also worth comparing the level of milk production that Planning Commission estimates had projected to provide 150 g of milk per capita for 660 million people by the year 1981, with the one cited by Whyte and Mathur (1968) to provide 170 g of milk per capita to entire population in the late 1960s. The Planning Commission figure for milk production to meet demand in the year 1981 was 36.42 million tonnes of milk per year, while Whyte and Mathur (1968) had cited the figure of 80 million tonnes of milk per year![20]

This detailed review of milk production targets in fourth plan has been attempted to underline that planners were not quite hopeful to achieve self-sufficiency sooner. This milk production targets, taking into consideration the milch animals, level of lactation yield and the availability of feed and fodder needed to push up production were studied by contemporary experts. In 1969, the surplus dairy commodity production in European countries appeared to offer a strategy and Kurien and his friend Michael Halse hit upon a plan. The idea was to supply reconstituted milk (by combining the donated milk powder and butter oil) and simultaneously build a national milch herd and a milk grid, by popularizing crossbred cow. The section below discusses the feed and fodder constraints by experts working out milk production targets and engages with the attention that grasslands and grazing got in expert speak.

8.3 Milk Production Targets and Feed Requirements in India

Talking about the issue of fodder and feed requirement, if India wished to achieve the projected milk production figures, R. O. Whyte (1967: 166), the then Grassland and Fodder Advisor to Government of India states:

> "The concentrates and oilcakes amount to about 10 million tons. The demand to feed the bovine population of India correctly is about eight times that amount[21]. The demand for work animals and poultry plus the bovines needed to supply to capacity (3 million litres per day) of the milk plants now (i.e. in 1966) in operation or under construction is six times the available supply. Data regarding the area of fodder crops and production are not reliable as these are not scheduled crops for statistical purposes in Panchayat records. The total production may be said to be around 150 million tons, assuming a yield per acre of 10 metric tons green, and the demand is about 600 million tons. Taking the average yield of straws and other residues from cereal crops at two thirds of a metric ton per acre, the amount available would be about 150 million tons to meet a requirement of 870 million tons."[22]

While Whyte notes that the availability of concentrates and oilcakes was one-eighth of the requirement, if India wished to feed its bovine population correctly, ironically even a portion of this was draining out of the country via export. Whyte (1967: 85) was aware of the volume of the export trade in oilcakes (mostly de-oiled groundnut cakes) and presents the figure of 840,000 tonnes (i.e. 8.4% of available concentrate) valued at Rs. 39 crores in 1964–'65 being exported per annum. Having acknowledged that "this would be equivalent to an indigenous production of 18 million litres of milk per day," Whyte (1967: 85) attempts to justify parting with this valuable input for milk production in the then prevailing environment. He does this by putting a rider for equating the quantum of export of oilcakes with an indigenous production of 18 million litres of milk per day, by stating that "if the other concentrate ingredients and green fodders were also available, and the livestock population was sufficient in number and productivity." Thus, Whyte (1967: 85) argues that if those conditions obtained, the amount of oilcakes that India exported could have added 18 million litres per day into milk production, however states Whyte, "since these conditions do not exist, *the oilcakes may as well be exported to earn the precious foreign exchange*"! A decade later and even after the launch of Operation Flood, the forage and fodder crisis continued unabated. National Commission on Agriculture (1976: 02) remarks that "only two-thirds of the fodder and one-fourth of the concentrates required for providing adequate nutrition to the present animal population are produced in the country"[23].

When it came to the discussion about animal nutrition during late 1960s, expert-think did not put any hopes on the "grazing grounds within daily walking distances from the villages". Pastoral and nomadic mobility that laid claims on grazing beyond daily walking distances probably did not even get a place in expert-think. Whyte (1968: 166–67) describes "the grazing grounds within daily walking distances from the villages" thus:

> With the present information and techniques that are available, it is difficult to assess the actual or potential contribution of natural grazing land to the overall feed resources for milk production. Village pastures and grazing grounds within daily walking distances from the villages are grossly overgrazed, sometimes carrying ten to twenty times the number of bovines that should be sustained by that particular type of grass community. On the other hand, there are vast areas inaccessible to cattle which should be used for the cutting of the hay. Although the species in the natural grass covers do acquire some degree of quality during the latter part of the monsoon months, they are for most of the year, low in protein and high in fibre, and as such of no value to productive milch animals.

Whyte (1968: 169) also appears to be giving a call for a rapid increase in the number of buffaloes in order to achieve the national target of producing 80 million litres per day, even with accounting for the reconstituted milk by relying on 16,00,000 tonnes of imported SMP, beside obtaining 32 million litres milk per day from 5,12,00,000 (51.2 million) buffaloes of breed able age inmilk and dry, giving the national average of 500 L per lactation. As per the 1961 Livestock Census, the population of buffaloes in the breed able age was 25 million.

Mathur and Knapp (1961: 343–344) had proposed that it would be relatively beneficial for a farmer to replace a pair of bullock for a tractor and utilize the fodder thus became available to feed two she buffaloes or one she buffalo and two heifers. Mathur and Knapp arrived at this conjecture by taking into calculation the quantitative equivalence between amounts of fodder required for maintenance of bullocks and milk animals on one hand and milk output on the other as reported in Farm Management Studies. The one they cite is, Gadgil and Gadgil (1940) and thus relevant only for a specific agro-climatic zone[24]. However, Mathur and Knapp (1961: 344) had gone ahead to project threefold increase in milk production from the then prevailing quantum of milk production under this assumption, if all 60 million bullocks more than three years old as per the 1956 Livestock Census in India were to get replaced by tractors, giving way for creating situation for equivalent rise in the numbers of female buffaloes. However, they had not discussed what such a policy advice would mean for resource poor farmers. As this study discussed in previous chapter and as it would recollect in the last chapter, such a policy advice was met with place-based and highly differentiated responses. So while in Ludhiana such a restructuring of bovine stock holding was taking place following the successful 'green' revolution practices there, Raipur did not show a fall in the number of bullocks. In Gujarat, despite the gradual increase of crossbred cattle numbers, indigenous male cattle population has not showed diminishing trend. The idea behind discussing all those contemporary proposals and policy advice in the sphere of dairy development strategies or on bovine stock holding and on the fodder question is to emphasise that the story of dairy development in India was much more complex than the simplistic 'replicate Anand pattern cooperative through intelligent use of dairy commodity aid' linear narrative.

8.4 Operation Flood Phase I (1970–1981)

While KDCMU had grown at rapid pace, we need to keep in mind, as pointed out by researchers from the 1980s onwards that during the 1946 to 1952 phase, on procurement front its expansion was thanks the gradual reduction in the share of Polsons. The procurement expansion was the function of village level milk cooperatives growing. Table 8.1 presents data on the number of primary milk cooperatives, members and milk procurement.

Similarly, on the distribution and marketing front, it saw the State-owned Aarey milk colony as competitor—since its access to Bombay Milk Scheme was mediated through this. KDCMPU had by then evolved a two-tier organizational structure—an organizational structure that dated back since the 1920s when Calcutta Milk Producers' Cooperative Union had emerged as an initiative to reckon with under the leadership of N. N. Bose. This book has discussed at length in an earlier chapter, that this two-tier organization structure was also recommended by a 1939 expert

Table 8.1 Growth of milk cooperatives at KDCMPU

Year	Number of villages societies	Members	Procurement in milk ('000 kg)
1955–'56	64	22,828	317
1960–'61	195	40,500	741
1965–'66	518	110,000	1370
1970–'71	706	180,000	3472
1975–'76	829	250,000	5292
1979–'80	895	308,000	7973

committee report on organizing cattle breeding and milk production on cooperative lines submitted to Government of Bombay. By the mid-1960s, when the Prime Minister Shastri called upon Varghese Kurien to replicate Anand, KDCMPU had an experience of having inspired and guided the establishment of similar cooperative efforts in two other districts in Gujarat, viz. Rajkot and Mehsana.

In the early 1950s, KDCMPU had accessed Skim Milk Powder under the PL 480 food aid programme, when its capacity to command a share in Bombay's fluid milk market had undergone a fluctuation due to Aarey milk colony having emerged as a competitor, and it planned to go into milk products in a big way. By 1960, KDCMPU had captured a large share of butter market from Polsons and had also started to manufacture cheese and baby food products. Although following prime minister's suggestion, National Dairy Development Board was set up in 1965, till the year 1968 we do not see it being mentioned in the contemporary literature at all[25]! The phase 1965–1970 in the history of NDDB is the least known phase, probably marred by many behind the scenes events arising from professional rivalry and state-owned dairy development corporations not quite willing to let go their control to cooperative structure, as claimed by Kurien's biographers[26].

However, during this phase, KDCMPU had already started to chart out the blueprint of what could be the new model of milk production in India. Government of Gujarat, which enrolled KDCMPU's technocracy to help with establishment of Rajkot and Mehsana Dairies, had also decided to set up three cattle feed plants to be taken over by cooperative unions eventually[27]. A 1967 project on dairy commodity aid of World Food Programme had supplied 45,885 million tonnes of maize and 24,391 million tonnes of sorghum for cattle feed, and 10,522 million tonnes of Skim Milk Powder at nominal rates of Rs. 350 per tonne[28].

Then around 1968–'69, there appeared some signs of hope for the National Dairy Development Board to replicate Anand pattern cooperatives, when it learnt about the mountains of milk powders and lakes of butter oil produced in an excessive surplus in European countries. NDDB proposed to use the donated dairy commodities in the form of 1.26 lakh skimmed milk powder and 42,000 tonnes of butter oil under the World Food Programme from the European Economic Community. Indian Dairy Corporation would transfer these dairy commodities to major city milk schemes and other approved dairy organization at approximate valuation of Rs. 4.35 and Rs. 9.67 per kg of skim milk powder and butter oil respectively. On the aforementioned values, the total proceeds to be generated were estimated at Rs. 95.40 crore[29]. The funds earned by marketing of reconstituted milk made from these donated commodities would feed into an eleven-point development programme to be carried out in 18 milk sheds spread over 57 districts and link milk producers there with four metropolitan cities. These milk shed areas extended mainly over 10 states located in Western and North-Western India, namely Punjab, Haryana, Rajasthan, Uttar Pradesh, Maharashtra and Union Territory of Delhi. Operation Flood became the largest dairy development project under the Fourth Five Year Plan and is even considered the world's biggest milk production drive. The first phase was initially planned to be of seven years' duration from 1970 to 77. However, the original duration was finally extended to eleven years (1970-1981).

Soon after the launch of Operation Flood Phase I, in its interim report on 'Milk Production through Small and Marginal Farmers and Agricultural Labourers', National Commission on Agriculture (1976) identified Operation Flood as a project which could be gainfully interlinked with other development schemes targeted to small and marginal farmers and agricultural labourers. It recommended that:

> In all areas of its operation for milk production enhancement, Operation Flood Project should invariably identify, induce and enroll all the small and marginal farmers and agricultural labourers for participation in its programme of milk generation, procurement and marketing. It should be ensured that at least one third of the producers in the project areas are drawn from the groups of small farmers and at least another one third from marginal farmers and/or agricultural labourers.

The commission underlined that Operation Flood implementation districts overlapped with twelve Small Farmers Development Agency (SFDA) project districts and eight Marginal Farmers and Agricultural Labourers (MFAL) Project districts. Thus, the commission sought it fit to recommend Operation Flood activities with SFDA and MFAL scheme. Commission also concurred with Operation Flood's thrust on crossbred cattle and suggested to extend crossbreeding by facilitating the replacement of indigenous cows by crossbred progenies. Towards this aim, the report stated: "The cattle development and milk production programme envisaged for assisting small and marginal farmers and agricultural labourers should aim primarily at replacement of the indigenous cows by crossbred progenies produced by farmers themselves or by supply of crossbred heifers and calves to them from other sources"[30].

However, switching from indigenous cattle to crossbred was not such an easy task. National Commission on Agriculture was aware that despite 'green' revolution strategy advocating to incorporate the production of green fodder, it had remained more of a lip service, as this thesis has discussed in last chapter. Even at Anand, the crop pattern had not taken to devoting larger part of farmland to fodder crops and relatively well-off milk producers had preferred to buy feed rather than grow more fodder. The output of green fodder even on irrigated land was nothing worth writing home about, and the area under fodder crops had remained more or less the same in last two decades. Talking about the demand supply gap on Feeds and Fodder Availability (see the following Table for figures), at another place in its report, National Commission on Agriculture observed, "This is a rather disturbing trend particularly in view of the fact that a number of important projects have been launched for development of animal husbandry" (Table 8.2).

Table 8.2 Feeds and fodder availability, 1973

	Concentrates	Dry fodder	Green fodder
Availability	11.0	309.0	214.0
Requirement	19.5	347.4	343.4
Shortage (%)	44.0	11.0	38.4

Source Report of the National Commission on Agriculture, 1976

8.5 Criticisms of Operation Flood Phase I

In the year 1973, a working paper by the Planning Commission suggested a livestock-based regional specialization for parts of Rajasthan and proposed to give dairy development a leading role. In an editorial for 'Economic and Political Weekly' (January 20, 1973) a concern was raised that such schemes that were proposed in past mentioned "provision of better grazing, feeding, watering, crossbreeding and marketing and transport facilities, but most of the programmes needed for raising milk production (or productivity of milch animals) remained on paper, while only the programmes for milk collection and its transport and marketing in the urban centres, which offers the prospects of immediate gains, and can be put through at relatively low cost and with little effort will be implemented"[31]. The editorial also voiced a concern that such lopsided dairy development was leading to serious back-wash effect in terms of deterioration in the diets of the rural dairy farmers, as the self-consumption of milk, by milk producers declined.

Even as there appeared celebratory reportage in media on the achievements of Operation Flood Phase I, in July 1978 Operation Flood authorities had to grudgingly admit that "milk schemes in metropolitan cities are likely to be disrupted because of the non-arrival of skimmed milk powder and butter oil from abroad" exposing the continued dependence of city milk schemes on the donated dairy commodities and imports. Despite the great influence that NDDB chairman wielded in policy circles, the authors of the Fifth Plan indicted Operation Flood for its "lack of emphasis on milk production programme"[32]. The Fifth Plan had stressed the importance of ending such dependence as early as possible. An editorial in 'Economic and Political Weekly' raised questions about the paucity of statistical data on milk production and stated that "The latest annual report of the Department of Agriculture claims that the average daily throughput of all the milk plants in the countries went up to 52.55 lakh litres in 1977–'78 from 44.40 lakh litres in the preceding year". That editorial wondered why "it is not disclosed how much of this increase was the result of larger availability of indigenous milk and how much was based on larger use of imported SMP and butter oil"[33]. Citing data from the annual report of the Department of Agriculture, this editorial also remarked that on average only 65% of the installed capacity of 84.87 lakh litres of all the milk plants was used in 1977–'78.

A critical editorial piece in the 'Economic and Political Weekly', following press conference in Bombay by the then chairman of Dairy Development Corporation, noted that India's dairy development policy was on the wrong tracks[34]. It criticized the DDC chairman stating that he had sought to discuss the progress of the scheme almost entirely in terms of the improvement in the supply of milk to four metropolitan cities, and worse kept harping on the improved availability of milk under the Bombay Milk Scheme particularly. The editorial indicted the Operation Flood for missing the targets consistently as in 1970–'71, the average daily throughput of milk of the four metropolitan dairies was 67% of target and this declined to 44% by 1972–'73.

This editorial mentioned that even the draft Fifth Five Year plan which found Operation Flood two years behind the schedule pointed out that the most unsatisfactory aspect is the lack of emphasis on the indigenous milk production programme. India's prominent weekly magazine that had carved out a genre called academic journalism was now suggesting that apart from probing the slow progress of milk production programmes there was a need to also investigate the larger economic and social implications of particular technologies and organizational forms for dairy development stressed upon under the scheme.

Singh (1979: 874–875) points out certain hard facts behind the EEC's largesse to third world country in the form of dairy commodity aid and expresses opinion that this was linked to promoting capital-intensive milk processing mechanism and through this towards increasing technological dependence in third world countries[35]. He links some lopsided policies on crop front that promoted cassava in Thailand and palm plantation in Malaysia, which ended up creating the feed exports from, and dairy surplus exports as food aid to third world countries. An Irish economist, Raymond Crotty (1980: 179) indicates the impact of Operation Flood-induced model of milk production on the landless in the following words:

> Competition from landeds' cattle forces the landless to abandon cattle keeping (only a relatively small proportion of the large landless class now keeps cows); or forces the landless to keep cattle on landed's terms for the fodder they must buy. That India's poor, who earns Rs 4 a day, must work for 25 days to buy a ton of straw, that is valueless elsewhere but costs Rs 100 a ton in India, to feed their cows, that traditionally fed on the public domain, denotes the grave retrogression for the country's largest, poorest social class. It is a situation that would seem to presage red revolution rather than "white revolution", as the envisaged explosion in milk production is referred to, somewhat glibly it might appear under the circumstances.

Around the end of the phase I of Operation Flood (after the extension was granted), 'Economic and Political Weekly' again found the programme having made India's dairy development a tale of rural–urban grid still. It reported that during that period just like the 'green revolution,' this development programme too had attracted its enthusiastic protagonists as well as bitter critics. This editorial was based on NDDB chairman's press conference to release the board's annual report for the year 1981-'82. In this press release, he claimed that annual milk production had increased from 22 Million tonnes to 33 Million tonnes[36] in a decade. Now if one would do a simple arithmetic, this meant effectively having the per capita milk consumption of 131 g, a gram below the level that was attained in the year 1951 and had gradually declined during the period 1951 to 1969–'70, bringing it down to 110 g during the Fourth Plan period[37]! Still while releasing the report, the chairman claimed that India had become 'self-sufficient' in milk consumption and milk products, barring cheese! Even in those days of intense scrutiny of Operation Flood by social scientists, no journalist seems to have asked Kurien why did he term that India had attained 'self-sufficiency' even as per capita milk consumption figures stood at the level attained in the year 1951, not even that editorial in 'Economic and Political Weekly'[38]!

The editorial found the excessive thrust of the press note on the so called 'national milk grid' disturbing, since while that phrase evokes an impression of widespread consumption and procurement, it actually denoted the flow from rural milk sheds to a relatively small number of cities with a concentration of purchasing power. The editorial voiced an opinion that "the effort to continually streamline milk marketing within this framework, must necessarily lead to a product pattern—milk powder, cheese, chocolates—calculated to meet the demand of the relatively well off in the cities and towns". While the editorial took note of the claim that increase in the procurement by four metropolitan dairies (from 7 lakh litres daily in 1970 to 13 lakh litres in 1981), and in other cities (from 3 lakh litres daily in 1970 to 10 lakh litres in 1981); it did ask a pertinent question: "To properly assess this achievement, it would have been useful to know how much of the increased procurement of milk came from the increased production of milk in the relevant catchment areas and how much from reconstitution of imported milk powder and butter oil".

Challenging the lofty claims of Operation Flood, Gandhian journalist Bharat Dogra (1982: 817) cited a note prepared by the Animal Husbandry Division on the estimates of milk production prepared for the triennium ending 1979–'80[39]. Quoting a detailed study by Indian Agricultural Statistics Research Institute (IASRI) of the impact of two milk supply schemes, Delhi and Mehsana for the triennium 1967–69 and then again for the triennium 1974–76, Dogra points out that there does not seem to be a significant increase either in milk production or in yield of milch animals. Dogra reminds us that according to targets of Operation Flood phase I, procurement by organized dairies in four metropolitan cities was to have increased from 6.87 lakh litres in 1970 to 22.5 lakh litres In 1975 (a threefold increase within five years?). He asks that if the milk production level had increased significantly, then these procurement targets should have been met, if not in 1975 then at least by the year 1981. However, as per a review of Operation Flood Phase I by UN experts, dairies in the four metropolitan cities could procure only 13.41 lakh litres in the year 1980–'81!

In a letter that *Economic and Political Weekly* published in its issue dated 2 April 1983, a well-known Irish livestock economist, Raymond Crotty underlined three adverse impacts of Operation Flood on poor landless livestock owners:

(a) It makes it more difficult for the majority of the villagers, who are poor, malnourished, cow less and landless, to buy milk, the only, or highly preferred, source of animal protein available to them;
(b) It is an incentive to add still more to the 18 percent of the world's bovines that are already crowded on to India's 3 percent of the world's land area, and thus force landless cow keepers to buy more fodder at higher price from landowners to maintain a given level of milk output and,
(c) It makes it more attractive for Indian landowners to divert resources from the production of food crops for the masses to the production of milk for the relatively wealthy in the cities.[40]

On 30 October 1983, Claude Alvarez penned a cover story article for *The Illustrated Weekly of India* titled 'The White Lie'. This article generated a great controversy provoking Varghese Kurien to publish a counter in the form of a booklet titled, 'The Black Lie'. However, the article by Claude Alvarez generated such a controversy that the government appointed an evaluation committee headed by L. K. Jha. In reply to a question in Rajya Sabha, the then Minister of State for Agriculture, Shri Yogendra Makwana stated that "it has been decided to evaluate the working of IDC/

NDDB as also the project implemented by them by a few experts"[41]. Promptly, 700 odd NDDB officials put in their resignations as a sign of protest. There was a similar protest and roadblock put forth by NDDB, when India's constitutional audit office, Comptroller and Auditor General of India wished to audit NDDB operations[42]. On 9 February 1984, the Government of India announced constitution of evaluation committee to be headed by L. K. Jha through an official gazette notification.

The State of India's Environment report (1984–85) published by Centre for Science and Environment observed that "since the 1970s, government's animal husbandry programmes have moved away rapidly from animal husbandry proper towards dairying, mainly to produce milk for urban areas. At a time of acute fodder shortage and given the vast majority of rural population which needs animals for ploughing, the government has no major programmes to solve this problem"[43]. Editors suggested that there is need for greater emphasis on animal husbandry, not milk production masquerading in the name of animal husbandry. They stressed upon employment to be secured for those who rely primarily on rearing cattle for a means of livelihood and suggested focusing upon the management of common lands for developing grazing resources. Editors had pointed out without mincing words that unfortunately government programmes have shied away from this crucial question, which involves all the thorny political problems of land ownership and management. Thus, we find that even as Operation Flood phase I was coming to an end, criticism against it was gathering steam. In 1984, Bharat Dogra also authored two small booklets titled 'Milk Imperialism: A Critique of Operation Flood Project' and 'Milk Muddle'. The critics of Operation Flood had underlined that contrary to the objective of making Indian dairy industry self-sustaining, the operation flood had ensured that India was more dependent on imports than ever. In an article published in Hindustan Times on 15 March 1986, Anjali Bhattacharya argued that "55% of the total milk output of Operation Flood dairy plants in New Delhi are made up from skimmed milk powder and butter oil. In Calcutta, this is 74%. Between 1979 and 1985, India received 20,000 tonnes of skimmed milk powder in excess of the earlier agreed amount. In 1984, India imported more than 25,000 tonnes of SMP from the USA. More than three quarters of this were commercial imports at dumping prices and the rest were donated."[44]

By this time, several researchers and scientists were pointing out the risks India would run if it continued to promote crossbreeding programme indiscriminately. It was being argued that the crossbreeding programme, if fully executed, will cause a considerable reduction in the number of suitable draught animals. When we look at 1966 and 1972 livestock census figures for Gujarat, we find that adult male cattle population was on a declining trend. Once more, India witnessed intellectuals and appropriate technology enthusiasts arguing that more attention should be paid to breeding dual purpose cattle, suitable as draught animals and more productive milk animals. Within two months after 'Hindustan Times' had published that article by Anjali Bhattacharya, Dr. Varghese Kurien gave interview to Mahesh Vijapurkar of *The Hindu* newspaper and sought to argue that critics of Operation Flood are like "the Ban the Bomb group". Kurien said that he resents being asked about demands made by people, whose credentials are not known[45]. Now, NDDB was seen arguing that “If crossbreeding has been resorted to, it has been selective, where non-de-script breeds have proliferated”. It was now being emphatically argued that crossbreeding has not been the NDDB thesis. It has merely followed the recommendations of the National Commission on Agriculture (1976) and this policy is older than NDDB! Those officials argued that they are following crossbreeding being an agency of the government and voiced a fear that conceding the demand for stopping crossbreeding would be taking away all the instruments for dairy development. They tried to state that dependence totally on crossbreeding alone is a slow maturing scheme and the NDDB has not made it a single point approach. But in 1984–’85, 13.3 lakh artificial inseminations were recorded. In their response to criticism on Operation Flood policies by social scientists, NDDB and Varghese Kurien appeared to be employing the dictum, *living in denial mode is the great defense.*

It would be nearly impossible to discuss all the critical writings on Operation Flood and the responses that came from NDDB. *Economic and Political Weekly* published many articles that were critical of the NDDB and also published in Letters to Editor section, responses from NDDB. However, the idea behind capturing some of the contemporary sources and discussion here in this book was to read it along the milk production targets discussion that this research engages in and reflect on the question if Operation Flood appeared to fail on providing the package of practices that would have provided an impetus on milk production related inputs, and raising the indigenous milk production with a pro-poor focus.

8.6 Operation Flood Phase II (1981–1985)

Immediately after Operation Flood II was announced, Singh (1979: 1765–1774) expressed concerns about the planning authorities’ neglect of assessing material inputs necessary for the successful roll out of the programme[46]. Even as his paper was based upon the feed and fodder requirement estimates for the crossbred cattle, he predicted:

> In case the shortage of feed and fodder is region-specific, there is every possibility that the gains of new technology of milk production may not reach the comparatively less irrigated and low productivity regions. Irrigation is a critical input for rearing livestock in general, and exotic milch animals in particular. In such a situation there is every possibility that the gains of crossbreeding technology will remain confined to some selected number of regions. It is likely that the existing regional disparities would be further aggravated when the programme is fully implemented. The possibility of more disparities within the region can also not be ruled out. Much will depend on the extent of adoption of this technology by small farmers, landless workers and big farmers in a region. It should be noted here that the gains of the recent breakthrough in crop production technology (HYVs) were confined to a few irrigated regions and the gains were comparatively more to big fanners. There is every possibility that those very farmers in those very regions would become the pioneers in adopting the new technology of milk production.

In addition to input constraints, Operation Flood Phase II was also criticized for having a selection bias. Singh and Sharma (1981: A 15–A 28) highlighted some concerns over the area choice in Operation Flood II Programme. They state that, "even as it has been claimed by project authorities that the selection of districts is guided by the 1972 report of the Dairy Development Subcommittee of the National Commission on Agriculture, this does not appear to be so in practice...It appears that in making the choice, much weight has been deliberately assigned to those states which are already more prosperous and more developed and have an in-built capacity for absorbing the programme"[47].

The programme had selected all the districts from Punjab and Haryana, while selecting 77% of districts belonging to Gujarat as well as Jammu and Kashmir as OF II districts. The authors used the data from Bhalla and Alagh (1979), popularly known as JNU-PPD study and demonstrated that the proportion of the districts selected for implementing the programme was smaller in the low productivity group and larger in the high productivity group[48]. They also observed that in general there is a tilt towards the high productivity and high growth rate areas in choice of districts. Singh and Sharma conclude the paper by stating that, "Since the new technology of milk production specified in the OF II, is strongly linked with irrigated land in the sense that the availability of feed and fodder is a crucial requirement, the inclusion of more districts from areas having higher level of irrigation and greater extent of HYV cultivation would further aggravate the situation of regional imbalances".

Huria and Achaya (1980: 1931–1942) underline some basic questions that still bear emphasis, despite strides made in dairy development, stating:

> The chief question is how to keep energy-output ratio as favourable as possible, in the *absence of sufficient pastureland* by which to reduce the cost of animal feeding. In turn this raises the question of whether grain and concentrate feeds will be available in sufficient quantity and at low enough cost to sustain the genetic potential of high yielding animals, and whether this can be achieved through total food system in which animals compete minimally with man for foods that both can consume. Another important question is that *how to preserve symbiosis between dairying and agriculture* (that has been all but lost in advanced countries) so as to provide a low cost energy system.[49]

The second phase was also getting launched at a time when National Commission on Agriculture had rallied behind the idea that Operational Flood strategy could help resolving the income disparities arising due to 'green' revolution interventions. In December 1971 itself, when not many evaluation studies of programme were available, an 'Interim Report of the National Commission on Agriculture' focused on the rhetorical promises of 'dairying as an instrument of social change'. The interim report paved the way for an official recognition of the celebratory narrative around Anand pattern and outlined how even union government's centrally sponsored scheme be dovetailed with Operation Flood. That report presented possibility of Small Farmers' Development Agency (SFDA) and Marginal Farmers and Agricultural Labourers (MFAL) funds being converged with OF. Five years later, when the final report of National Commission on Agriculture got published, it presented the worrisome picture on the front of pastures, grazing lands, feed and fodder resources. The planners of Operation Flood Phase II, instead of taking the programme to rain fed areas with history of contributing to dairying, further accentuated the position of landed farmers in the beneficiary zone of 'green' revolution and left out those one that remained in the rain shadow of green revolution[50].

Nair and Jackson (1981:2219–2232) presented an elaborate alternative to Operation Flood Phase II strategy which argued to walk a different path and alerted that the officially anointed strategy would make the question of animal nutrition even more pronounced. It presented an argument that local feed and fodder resources would also now start flowing out (just as milk has kept flowing upward) and end up further accelerating the fodder scarcity. The feed plants had started to change the rhythm of milk production from being a pastoral and rural based to purchased inputs based, where landless poor bovine stock rearer stood chances to get further discriminated against. Irony was that NDDB experts did not bit an eyelid to tell landless what its commitment were to "degraded *gauchar* (grazing land)," but the very cattle feed plant whose moment of inauguration witnessed the call for replication of Anand pattern was located on a land that was once a grazing land supporting cattle and buffaloes of nearby villages!

8.7 EEC Milk Out of India Campaign (1985–1988)

India Committee of the Netherlands launched a campaign in the year 1985 titled, EEC Milk out of India[51]. This campaign aimed to halt, what it called 'the vicious circle of EEC dairy aid to, and animal feed imports from, India'. Government of India had requested European Economic Community (EEC) for dairy aid from 1986 to 1990 and a decision was expected in the early 1986 (sometime in March). The campaign was directed at the European parliament, the European Commission (the 'executive board' of the EEC), the parliaments and governments of the EC-member countries and the compound feed industries in these countries. The campaign had published a short booklet in Dutch language titled, 'India as EEC Milk Cow' and had released a campaign manifesto[52]. Under this campaign, researchers made four demands on the policy makers at EEC:

1. Phase out dairy aid to India within two years.
2. No dairy aid for bottle feeding.
3. No aid for exotic crossbreeding in India.
4. Stop EEC animal feed imports from India.

The European Commission officials prepared yet another report in early 1986. The content of this report resembled more or less the style that was there in the previous mid-1985 report by EEC. Commenting on this report, Malcolm Subhan wrote in the journal 'Commerce':

> It was based more on information available with the Commission than recent field work by its officials. It concluded that as a development project Operation Flood had given good results so far although, like any human endeavour, there was room for improvement. The report agreed that Operation Flood should be continued, to allow its objectives to be reached, but noted that this did not necessarily imply continued EEC involvement through the supply of SMP and butter oil. Such aid could only be justified if it could be shown that domestic production had not increased to the point where it could meet the growing demand for milk.

Following this, in April 1986, the general assembly of European Development NGOs adopted a resolution on the subject of continued EEC support for Operation Flood III. The resolution urged an urgently needed reorientation of Operation Flood in favour of rural and urban poor.

In a news report published in Economic Times dated 6 December 1986, Brussels-based correspondent voiced a conjecture that EEC food aid for Operation Flood III can be ruled out and attributed this 'news' to senior EEC officials[53]. EEC Consultants' report, released sometime around December 1986, put forward the following suggestions on the input services programme within Operation Flood:

> Genetic upgrading of milk animals should be oriented towards Indigenous breeds which, within specific ecological regions, have proved to have milk potential, and towards buffaloes. Specific research efforts should be directed at dealing effectively with seasonality in breeding. Crossbreeding programmes should be implemented only in those areas, where climatic conditions are suitable and adequate fodder and good veterinary infrastructure are available.[54]

The joint review mission also underlined the need to take care of "the social stratification of producers and their level of institutional organization" so as to "minimize the inevitable unequalizing income effects which accompany major technological changes". The mission report cautioned Operation Flood authorities that "large farmers, with greater capital resources and ability to bear higher risk, may derive undesirable advantages from the technical change". The review mission did not approve of long distance transport of liquid milk as a practice that could potentially become and everyday reality and could turn out to be a trigger for an unfair competition with local producers, when it stated, "As a general criterion, milk should be produced as near as possible to the centres of demand. Long distance transport of liquid milk, now an established and somewhat necessary practice, should be reexamined, taking into account open and indirect subsidies which make the procurement from long distances not only technically but also economically a viable proposition. Special care should be taken so as to avoid unfair competition with local producers."[55]

The joint review mission of EEC and World Bank which evaluated the implementation of Operation Flood II came to the conclusion that "a moratorium on the imports of dairy products is more than appropriate and deliveries of milk powder and butter oil should take place only if and when a substantial and documented deficit shows up". Mission put forth an opinion without mincing words that "imports of dairy products (either donated, subsidized or commercial) can, in the present situation of growing indigenous stocks, damage the strengthening of the cooperative industry". On the level of differential achievements of targets, the report stated, "The quality of achievement is quite uneven across India and the states, ranging from satisfactory to disappointing. The situation has become very complex and very demanding in terms of rationalizing present operations and building on them"[56]. Review mission had also argued that "Technical and economic complementarity between crop and livestock activities calls for a 'farm system' analysis and monitoring programme, to date absent both at project and government level". From the available statistics, they also noted that "The feed component in cash expenses and in the total cost of milk production is fairly high and is proportionally higher for households with little or no crop land from which to obtain roughages and crop residues".

Following the release of the report by the joint review mission of the World Bank and EEC, Claude Alvarez wrote another critical opinion for 'The Illustrated Weekly of India' titled 'Body-blow' wherein he indicted even the evaluators for not speaking out against Operation Flood earlier. Alvarez also pointed to "an internal contradiction within the state politics of Maharashtra that brought up fresh spurt of opposition to Operation Flood". Alvarez talked about "an internal note that the Maharashtra milk department had put up before the Jha committee arguing that Operation Flood dairies set up in Jalgaon and Kolhapur had been miserable failure". Pointing out what had upset the milk officials, Alvarez stated that "their own milk powder was being discriminated against by Indian Dairy Corporation, which had developed a vested interest in selling imported powder gifted to it freely by the EEC"[57].

Then on 20 January 1988, in the wake of the European Parliament discussing the support to Operation Flood phase III, India Committee on the Netherlands issued an urgent call upon the European Parliament reiterating the demands of the campaign 'EEC Milk out of India'. However, in March 1988, the European Commission decided to support the third phase of Operation Flood with new multiannual food aid supplies, despite years of debates and controversy in India and Europe. The campaign EEC Milk out of India seems to have achieved small victories though, since the EEC decision to continue to support third phase of Operation Flood included certain special conditions that were agreed upon by the Indian authorities. These conditions included, i) pricing of food aid supplies at the same level as indigenous milk price, ii) annual review to take place of the requirements of commodities and shall inform allocation of quantities from the overall multiannual commitment of 75,000 tonne of SMP and 25,000 tonne of butter oil and iii) food aid to be exclusively utilized for recombination into liquid milk and any other use would have to seek prior approval of the European Commission. In addition to this special conditions also provided for i) system of reviewing annual plans jointly by the project authorities, the World Bank and the CEC; ii) periodical progress reports to be submitted to CEC by project authorities and iii) access to all project related documents and inputs about CEC financing to representatives of the Commission and the Court of Auditors of the European Communities and visiting missions. Last but not the least, the special conditions also stipulated continued EC participation in Operation Flood phase III, subject to the signature of the loan agreement between India and the World Bank[58].

8.8 Operation Flood Phase III (1987–1994)

This third phase was planned as a phase of consolidation of the network of Anand pattern cooperatives further with a target of creating 70,000 village level milk cooperatives, with 15 million milch animals. By 1994, Operation Flood had aimed to have covered over 176 milk sheds and to supply 13 million litres per day to about 450 towns and cities. As against this aim, at the end of 1993, some 65,300 village level milk cooperatives were functioning in 170 milk sheds in the country involving 8.40 million milk producers.

The tenor of social science literature on Operation Flood had started to change slowly, but subtly by the end of 1980s. Academicians who looked to the promise of the programme were showing the signs of irritation and were now seen arguing that social science researchers who held critical views on Operation Flood were ideologically predisposed to criticize the programme like those who were raising questions on Narmada Dam[59]. The common ground for dialogic reflection and setting the course correction appeared to be under the risk of getting irretrievably lost. Parthsarathy (1991: A 181) used NSSO data on the land and livestock correlation and put the rhetorical claim that Operation Flood programme has been scale neutral and milch animals holding shows more egalitarian distribution compared to land ownership to rigorous scrutiny. He concludes Operation Flood's bias towards landed milk producers by stating,

> Firstly, landless labourers do not benefit from milk production directly due to the fodder barrier. Therefore, the claim that milk production has the potential to benefit the poor is exaggerated. Secondly, even among marginal and small owners more than around two-thirds at the all-India level do not participate in milk production because they have credit and other barriers. Thirdly, the rest, i.e. around one-third who participate in milk production do not appear to suffer either in relation to the quality of cattle maintained or in relation to milk yield. Fourthly, the dairy cooperatives are not immune from the local power structure. Fifthly, despite these limitations, the distribution of income from milk production is much less unequal than the distribution of income from crop production.[60]

However, the third phase of Operation Flood where the crossbred cows had started to show rapid rate of diffusion also occurred when the cooperative sector within milk economy had started to lose the protective support under the neoliberal shift in government policies. In 1996, as the third phase of Operation Flood came to a close, World Bank did not renew support for cooperative dairying programme. Candler and Kumar (1998: 12) quotes a 1995 internal Bank Memorandum that viewed private sector as potentially more efficient than cooperative sector and underlined that continuance of Operation Flood was too costly to support. India's Tenth Five Year Plan also continued to support and reiterate this view. If one were to argue on the basis of a mere cursory glance at the milk production figures in India post the neoliberal shift, one may come to believe that privatization has not resulted in deacceleration of the growth.

However, animal genetics scientists in the decade of 1990s started to study the crossbreeding and argued that per animal milk productivity figures still leaves much to be desired and indiscriminate diffusion of crossbreeding has created its own problems. There are few who have argued for the need for second Operation Flood type of intervention in recent times and World Bank has now revised its thinking and came out to support a very ambitious National Dairy Sector Programme loan for the period 2011–2017. This research does not look into this ongoing World Bank assisted dairy development programme, since our objective was to look at the 'green' and 'white' revolution period. In the next section, there is a discussion on what was happening on the state of grassland and grazing.

8.9 Returning to Grasslands and Grazing

Uttari Rajasthan Sahakari Dugdh Utpadak Sangh Limited (URMUL) was established in the arid Thar region of Bikaner, Sriganga Nagar and Churu districts on 28 August 1972, as part of Operation Flood Phase II. In the next twelve years, this cooperative grew leaps and bounds and while at the start milk procurement was merely 200 L per day, in the year 1984 it had reached 1,50,000 L per day. A decade later, Ghose (1994) observed that people from villages (Rajpuria and Nathusar) around Bikaner now had tendency to move to access the grasslands of Jaisalmer, where *sevan*—the best grass in this region—was ready to be harvested. Ghose however, remarked that this survival strategy cannot last beyond a couple of years since the grasslands of Jaisalmer were being rapidly colonized by the government under Indira Gandhi Nahar Pariyojna. In his article written during a drought year of 1994, he had warned that "soon enough, the grass will disappear, like it has in Lunkaransar and Bikaner".

Ghose's narrative presents the community management of grazing land in villages as a thing of past that is recollected in people's recounting of those old days. Sanjay Ghose, a well-known Indian social worker, notes the shift when he writes:

> Now, state services have diminished any sense of community ownership. What has become of Kubiya's *Gauchar* is symptomatic. Some of it has been occupied; every year the fence of the common pastures are driven inwards by illegally expanding private holdings. Everybody's property means nobody's property and the first victim has been self-sufficiency. The vast majority of villages now depend either on residual crops from individual farms or the market.[61]

A few years later, in April 1999, in a cover story published in environment fortnightly, 'Down To Earth' described without mincing words how "misguided development efforts are destroying India's grasslands". This report cited Indira Gandhi can as a project that had "adversely affected the *sewan* grasslands, which have provided highly nutritious fodder for cattle in Thar desert over the centuries"[62].

By the early 1990s, the flood of milk did not merely look up to the small streams of small and marginal farmer member of dairy cooperatives. By now, dairy development was not merely an instrument of social change, but it had evolved as heavily technologized sector that revolved around the idea of enhancing milk yields. In 1992, the National Dairy Development Board had planned to import 600,000 doses of *Bovine Somatotropin* (BST), a synthetic growth hormone, which was banned by the European Commission, pending the outcome of tests of its long-term effects on cattle. However, the proposal ran into stumbling block, when Indian Council of Agricultural Research raised objection against large scale import of what it termed as "a new chemical that has not been tested for its suitability to India"[63]. However, two years later, a team of scientists led by R. S. Ludri of the National Dairy Research Institute (NDRI) Karnal had carried out experiments to inject non-pregnant/ dry crossbred as well as indigenous cows and buffaloes with progesterone oestrogen hormones. They had also started to take their technique outside laboratory and were reported to have injected 20 animals reared by farmers around Karnal[64]. While Ludri and his team were carrying out experiments to tackle breeding irregularities, there existed a real risk of it being used indiscriminately by veterinary practitioners without paying due attention to long-term impacts.

In April 1994, two scientists of the National Dairy Research Institute, Karnal (R. S. Singh and Prakash Mohanty) reported following a survey of Delhi Milk Scheme's bottled and powder milk that it contained 0.50 and 1.01 ppm of DDT residues. Similarly, a nationwide survey carried out by Indian Council of Medical Research (ICMR) had also established that the residues of chemicals and pesticides in milk and milk products had crossed safe limits[65]. Towards late 1990s, even as the Operation Flood planners were showing a graphic presenting of steep upward trend for indigenous milk production, there had also emerged another discourse. This discourse not only termed India's monsoonal grasslands as poor but underlined the diminishing carrying capacity of the same. The reports in scientific journals around late 1990s suggested that "the state of Kerala has grasslands with highest carrying capacity, whereas Haryana—the cradle of 'Green' revolution in 1960s—now reported grasslands with lowest carrying capacity, around one seventh compared to Kerala"[66]. By now N.S. Jodha's research on the Common Property Resources had appeared and made the research community aware on the issues around the management of commons. While the discourse on 'carrying capacity' of grasslands at times ran parallel to the 'protected area' discourse, being laden with conservationist terminologies, Asad Rahmani, the then director of Bombay Natural History Society, stated "There is no pure grassland sanctuary or park where all the elements of a grass ecosystem can be studied"[67]. Rahmani pointed at the 'protected area' identification process which was exclusively around "according protection to an area for its flora and fauna". However, this suggestion by Rahmani of the absence of a purely grassland sanctuary must remind us of colonial animal husbandry expert, F. Ware (1938: 11) suggesting at the second Cattle Conference in 1937, "sanctuaries should be formed for domestic animals in forest areas".

An article that formed part of 'Down To Earth' cover story on the state of grasslands (15 April 1999) listed the following policy changes:

> a, Government stress on quality of livestock, rather than quantity. b, Overgrazing may be an important reason for degradation but the onus really rests with institutional policies rather than the livestock owners. c, Long term strategy based on sound research needs to be chalked out to meet production targets with fewer animals. d, There is no rangeland policy. As things stand the Union ministry of environment and forests and the ministry of agriculture are quibbling over whose baby, grazing really is. Existing policies for pastoral areas fail to appreciate the practical or valuable implications of traditional pastoral systems. These systems should be researched objectively and improved upon to meet current requirements. e, National policy on livestock population and the creation of forage resources has to be formulated -- and implemented. Public support will come forth only if some alternative arrangement to meet grazing requirements can be suggested. f, Before any programme for rangeland development is even considered, it is essential that local needs be integrated into the science of it. g, One of the biggest gaps is the absolute lack of data and surveys. This situation needs to be corrected on priority.

Thus, despite the three phases of Operation Flood, the livestock sector in India—and the Dairy Development agenda in particular—continued to remain locked into capitalist mode of production system, somewhat oblivious to landless and marginal farmers' concerns. Indiscriminate diffusion of crossbreeding during Operation Flood also created its own problems. Gandhian organization Akhil Bharatiya Krishi Gauseva Sangh (All India Association for Agriculture and Cow Protection) had published in the year 1981 a report of an expert committee that was constituted in June 1979 to study exotic crossbreeding in India. The expert committee under the leadership of Dr. M. Y. Mangrulkar, a retired principal of Jabalpur veterinary college and ex-director of Madhya Pradesh Animal Husbandry Department visited sixteen states throughout India to understand the effects of exotic crossbreeding. The committee had submitted its report in November 1980, where it argued that there had been a progressive depletion of the draught animal power. The report had also showed that male calves born by exotic breeding were mostly given over to butchers and those who survived were often too weak to stand the strain and climatic conditions.

In 1997, Dr. Anil E. Nivsarkar, former director of the National Bureau of Animal Genetic Resources, had expressed his dissatisfaction with what he termed 'indiscriminate crossbreeding' in an interview with environmental journalist, Indira Khurana. Dr. Nivsarkar stated that initially, crossbreeding with Jeresy and Holstein germ plasm was to be undertaken on *non-de-script* cattle to improve their performance. But the programme was actually implemented using well-defined breeds, whose performance was satisfactory anyway. Posing a question on the benefits of such indiscriminate crossbreeding, he points out that in fact, successive generations of crossbred cattle have shown a decline in milk yield[68]. That article in 'Down To Earth' had pointed out that Red Sindhi, Sahiwal and Punganur cattle breeds are heading towards extinction. Similarly, in 2013, Jay Mazoomdar penned an alarmist article in the weekly magazine 'Tehelka' painting a sad tale of almost extinct breeds of indigenous cattle[69].

In early 1990s, under the liberalization regime, the State took away the protective measures for dairy sector and opened it up for private sector participation by making changes in the law that regulated essential commodities and specifically Milk and Milk Products Order. By the year 2014, the private players had commanded so much of milk markets in India that it surpassed the market share of cooperatives. NDDB's annual report for the year 2010–'11 grudgingly conceded, "It is estimated that the capacity created by them (private dairies) in the past 15 years equals that set up by cooperatives in over 30 years". A cover story article in environmental fortnightly, *Down To Earth* recently argued that "While the private (dairy) companies are surging ahead, (dairy) cooperatives have been mired in politics and shady dealings. Several cooperatives are now lying defunct, though animal husbandry department is yet to do any comprehensive study on this"[70].

The next chapter puts on record a few voices from contemporary animal science and technology experts, who have underlined the need to reintegrate crop and cattle on rural India's agrarian landscape. It also presents an analysis of the changes in bovine population in Gujarat for the period (1997–2012) and for Kheda (1987–2012). It also recollects some insights on changing mode of production by milk cooperative members who now keep large size herds.

Notes

1. These dairy technocrats were namely Varghese Kurien and H. M. Dallya, who worked as Manager and Assistant Manager at KDCMPU.
2. Editorial, *The Economic Weekly*, 29th October 1955.
3. Rao, G. A. (1956) 'Milk for Millions,' *Economic and Political Weekly*, (May 19), pp. 585–587.
4. This document was circulated widely amongst those whose voices counted in deciding dairy development policy in India. For instance, the copy that I was able to locate at Sectoral Analysis and Studies division of the National Dairy Development Board, Anand was the one presented by the then Asst Manager of KDCMPU on 25th December 1958 to Dr. Burde, who was then a lecturer in Dairy Development and Dairy Administration at National Dairy Research Institute, Karnal.
5. Philipose, V. A. quoted in Singh, S. P. and P. L. Kelly (1981) *AMUL: An Experiment in Rural Economic Development*, Macmillan, New Delhi, p. 39.
6. Khurody, Dara N. (1974) *Dairying in India: A Review*, Asia Publishing House, Bombay.
7. As we would discuss, the problem of city cattle stables have remained as a constant presence, despite the discourse on how the milk production based on city cattle stables is unsustainable and uneconomic proposition. During the Operation Flood years, it was proposed by NDDB to spend a part of project funds towards the programs for rehabilitation of city cattle stables, but the efforts have not met with success.
8. By popular narratives, one has in mind biographies that borders on being hagiographies on Varghese Kurien, Kamath, M V (1996) *Milkman from Anand: the Story of Varghese Kurien*, Kurien, Varghese (as told to Gauri Salvi) (2005) *I too had a Dream*, Roli Books. These are the narratives that rely largely on long personal interviews with Kurien and his colleagues in KDCMPU and NDDB.
9. At the occasion of a golden jubilee of Indian Dairy Association, Dara N. Khurody was asked to articulate his views on *future measures needed for development of Indian Dairying*. Somewhat bitter sounding, Khurody recounted that "In February 1971, I was asked to present a paper on 'Consolidation of Indian Dairying' at the 8th Dairy Industry Conference at Trivendram, Kerala, in which the history of Indian Dairying in the form of jottings, since the turn of the century was briefly narrated. That paper also dealt with certain measures needed for consolidation. Like many other reports, before and after, by individuals, committees and commissions, it may also be gathering dust somewhere. This, however, should not discourage anyone from expressing his ideas. Indeed, all actions begin with thoughts". As he went on to narrate in his preface, the paper he wrote on 2nd October 1973 dealt with subjects like, "(1) Need for organising the different sectors of dairy industry so that they could play their rightful role in future (2) Need for stoppage of non-technical men, from being head of Dairy Departments (3) The lacunae with the composition of NDDB and IDC (4) The XIX International Dairy Congress, 1974 (5) Need for setting up of a National Dairy Committee (6) Milk Colonization projects and the role they should play (7) The NDRI and its future (8) Indian Dairy Science Association and what it should do (9) Dairy Technical Staff (10) Stringent measures for prevention of adulteration of milk and milk products and compulsory distribution of pasteurised milk (11) Production of more milk—the most pressing need of all (12) Dairy Statistics (13) Foreign Aid Policy (14) The problem of surplus cattle (15) Milk and Milk Product Substitutes etc.
10. Editorial, *Economic Weekly*, 29th October 1960.
11. For more details on Aarey milk colony and FAO partnership, see papers archived at New Zeland Hostel, Aarey Milk Colony, Bombay. See the news item published in the journal of American Dairy Science Association titled 'FAO report on Bombay, India'.
12. It is curious to find such a strong worded criticism of a strategy to develop dairying in India through dairy commodity aid that came from the surplus production in Europe very early on in mid-50s, when such a foreign assistance was being used to strengthen milk colonies in peri-urban areas in major metropolitan cities. However, the same strategy to 'replicate Anand' becomes 'a visionary use of commodity aid' during Operation Flood years in the views of Kurien and NDDB in 1980s!

13. For a detailed discussion of the implication of FAO assisted dairy development projects in that era of City Milk Schemes, see 'Calcutta Milk Scheme under FAO Auspices', in *The Economic Weekly*, September 15, 1956. Pp. 1088–1089.
14. Chapter on 'Animal Husbandry,' *National Commission on Agriculture*, Brief Overall Report, p. 308.
15. As per an estimate only about 110 g of milk per capita per day was available during the Fourth Plan period, a decrease from per capita milk consumption at 132 g during the year 1951. See, Parthsarathy, G. (1991) 'White Revolution, Dairy Cooperatives and Weaker Sections,' *Economic and Political Weekly*, Vol 26, No 52 (December 28, 1991), p. A-178.
16. Bhattacharya, P (1951) Note on 'The Possibilities of Introducing Artificial Insemination on a Field Scale in Suitable Areas', Note prepared for the eighth meeting of Animal Husbandry wing of the Board of Agriculture and Animal Husbandry in India held at Mysore on 17 to 22 February 1949, Government of India Press, Simla, pp. 160–164.
17. Proceeding of a meeting on 8th January 1952, where Dr N C Wright presented his paper, 'Some Problems of Tropical Dairying,' noted that "India and Pakistan, together produced over 6000 million gallons a year – a figure about three times as high as that of the United Kingdom".
18. Chatterjee, I (1960) 'Milk Production in India,' *Economic Weekly*, (September 03, 1960), pp. 1347–48.
19. See, Sukhatme, P. V. (1970) 'Nature of Protein Problem and its Implications for Policy Measures,' Paper presented at National Food Congress, New Delhi.
20. National Commission on Agriculture (1976) projections on daily per capita and gross demands for milk had put the figures of milk production required to meet the same at 44.17 million tonnes by 1985 and 64.40 million tonnes by 2000 and added that "it may appear to be well-nigh impossible to achieve this much increase in milk production"!
21. At other place in the same book, Whyte (1968: 84) states that "To achieve 80 million litres per day of milk production (6 oz. or 0.17 L per head of human population per day), we need 89,600,000 tons of concentrate ingredients (assuming that sufficient green fodder and crop residues are available to provide a balanced ration) to feed 256 million animals with a lactation yield or herd average of 500 L. With lactation yield of 1000 L per animal, we need 128 million animals and 51,200,000 tons of concentrates. With a lactation yield of 2500 L per animal, we need 51.2 million animals and 33,280,000 tons of concentrates, or one-fifth of the animals and about one-third of the amount of concentrates to produce the same total amount of milk.".
22. At other place in the same book, Whyte (1968: 83) refers to the state of milk pants that were under operation in late 1960s and states that those were working at one-third of their capacity.
23. *National Commission on Agriculture,* Part VII (Animal Husbandry), Ministry of Agriculture and Irrigation, Government of India, New Delhi, 1976, p. 02.
24. Gadgil D. R. and V. R. Gadgil (1940) *Survey of the Wai Taluka,* Gokhale Institute of Politics and Economics, Poona.
25. R O Whyte's 1968 book *Land, Livestock and Human Nutrition* for example doesn't list NDDB in index or anywhere throughout the discussions on milk production targets and how to achieve them!
26. NDDB's popular propaganda literature as reproduced in M V Kamath's *Milkman from Anand* does give such an impression, see pp. 150–152.
27. Few figures from the Livestock census in Gujarat states would be worth perusing to see how KDCMPU effects were spreading to other districts that offered promise to replicate the *real Anand pattern* (read: buffaloisation). While Kheda reported the largest population of buffaloes, it was closely followed by Mehsana and together these two districts accounted for 28% of the buffaloes in the state. The population of she-buffaloes ranged between 1.50 and 1.75 lakhs in Panchmahal, Banaskantha and Sabarkantha; and it was 1.25 and 1.50 lakhs in Ahmedabad, Junagadh and Vadodara. The ninth on that list having higher number of she-buffalo population was Bhavnagar which recorded 1.07 lakh heads of she-buffaloes. See, Report and Provision Figures of 11th Livestock Census for Gujarat State (1972), Bureau of Economics and Statistics, Government of Gujarat, Ahmedabad. p. 09. In 1972, even the provisional figures of Livestock census had shown that there was a decline in the number of male cattle by 1.5% compared to

the number of male cattle recorded in 1966 census. Thus, the onset of Operation Flood had triggered the process of decline of the number of plough bullocks very early on in Gujarat, a trend that was set to get further accentuated during Operation Flood II.

28. Alvarez, Claude (1985) 'Imperialism through Food Aid: The Role of Third World Elites,' in Alvarez, Claude (ed) *Another Revolution Fails*, Ajanta Publications, New Delhi, p. 03.
29. See *Interim Report of the National Commission on Agriculture on Milk Production through Small and Marginal Farmers and Agricultural Labourers*, Ministry of Agriculture, Government of India, New Delhi, December 1971, p. 25.
30. Ibid p. 02–03.
31. See, 'Draining the Villages?' *Economic and Political Weekly*, January 20, 1973; p. 89.
32. See, The Fifth Plan document for more details.
33. Editorial (1978) 'Operation Flood: Time for a Closer Look,' *Economic and Political Weekly*, July 15, 1978; p. 1122–23.
34. Editorial () 'Dairy Development: On the Wrong Tracks,' *Economic and Political Weekly*, p. 457.
35. Singh, Narendra (1979) 'Unloading White Revolution,' *Economic and Political Weekly*, (May 19), pp. 874–75.
36. Compared to this, in the year 1979–'80 the milk production figure was pegged at 30.2 million tons. Even as Kurien claimed with square face that India has attained self-sufficiency, the figure of milk production turned out to be more than 10 million tons below the level projected to be achieved by the year 1985, as per demand projections reported in National Commission Agriculture (1976)!
37. As we have noted earlier, a careful scrutiny of the per capita milk consumption figures over the years does indeed lead us to recognize that in 1951, the per capita milk consumption stood at 132 g.
38. Kurien sought to ward off criticism in his letter to editor (April 02, 1983) by defining 'self-sufficiency' in milk and milk products as, "All that we can say is that, at the present time and at the present prices, the demand for milk and milk products is being met". Did Kurien not bother to see the milk production targets at the start of Fourth five year plan that appear to have taken a minimum nutritional requirement of 201 g per capita in terms of fluid milk to attain 'self-sufficiency'? See, Kurien, Varghese (1983) 'Letter to Editor,' *Economic and Political Weekly*, (April 02).
39. Dogra, Bharat (1982) 'Milk Production: Not Such a Flood,' *Economic and Political Weekly*, (May 15), p. 817.
40. Crotty, Raymond (1983) 'EEC Interest in Operation Flood,' *Economic and Political Weekly*, 2nd April 1983, p. 522.
41. See, proceedings during Question Hour in Rajya Sabha, 2nd December 1983.
42. In Kurien's biography, *I too had a Dream*, he revisits that episode and tells his readers that it was also a case of personal ego and procedural shortcuts to victimize him. NDDB had taken the matter to court in the year 1998, challenging the CAG of India getting presidential approval to audit NDB through Writ Petition 4834 of 1998, and many years after on 27th January 2010, the High Court of Delhi had ruled in favour of CAG of India, upholding constitutional auditor's right to audit a public sector body constituted by an act of parliament.
43. See, 'Milking the Land', in *State of India's Environment* 1984–85, Centre for Science and Environment, New Delhi, pp. 15–16.
44. See Bhatacharya, Anjali (1986) 'EEC Dairy Aid eroding Self-Reliance Efforts', *Hindustan Times*, 15th March 1986. http://www.indianet.nl/a860315.html
45. Vijapurkar, Mahesh (1986) 'Commodity Aid for Dairying: Threat that was converted into Opportunity', *The Hindu*, 13th May 1986. http://www.indianet.nl/a860513.html
46. Singh, Surendar (1979) 'Operation Flood II: Some Constraints and Implications,' *Economic and Political Weekly*, October 27. Also see, Nair, K N (1980) 'Operation Flood: Some Constraints and Implications,' *Economic and Political Weekly*, February 23.
47. Singh, Surenddar and R K Sharma (1981) 'Some Implications of Area Choice in Operation Flood II Programme,' *Economic and Political Weekly*, March.

48. Bhalla, G S and Y K Alagh (1979) *Performance of Indian Agriculture: A District wide Study*, Sterling Publishers, New Delhi.
49. Huria, Vinod and K T Acharya (1980) 'Dairy Development in India: Some Critical Issues,' *Economic and Political Weekly*, Vol 15, No 45–46, pp. 1933–1942.
50. Kapur, Depinder et al. (2012) *A Commons Story: In the Rain Shadow of Green Revolution*, Foundation for Ecological Security, Anand https://www.nabard.org/Publication/OP_60_commons_story.pdf
51. To be precise this campaign was launched on 9th December 1985. See the press release at the time of launch of the campaign here, http://www.indianet.nl/pb851209.html
52. For the full text of the campaign manifesto, see http://www.indianet.nl/milkman.html
53. An archived copy of the newsclip can be accessed here: http://www.indianet.nl/a861206.html
54. For the excerpts from the Review Mission's report, see http://www.indianet.nl/pb8703i.html
55. Many years from that moment when such long distance transfers of liquid milk becoming and everyday reality and leading to unfair competition with local producers, during the year 2015 researchers points out that Amul has precisely indulged in such a practice in an effort to capture Hyderabad city's milk market. Sagari Ramdas argues that "Amul has begun to do to the informal dairy sector what the European Union threatened to do to the Indian dairy sector: dump milk and milk products, capture the market and then drive down procurement prices as well". For details see, Ramdas, Sagari R. (2015) Death of Small-Farmer Dairies amidst India's Dairy Boom', *Economic and Political Weekly*, Vol L, No 9, pp. 21–23.
56. For more details see Press Release by the campaign EEC Milk Out of India, India Committee on Netherlands, March 1987. http://www.indianet.nl/pb8703.html
57. Alvarez, Claude (1987) 'Body-blow', *Illustrated Weekly of India*, 19th July 1987. http://www.indianet.nl/a870719.html
58. This is paraphrased for the sake of brevity. See the exact language of the special conditions from the Draft Commission Decision allocating food aid to India archived here: http://www.indianet.nl/pb8803-1.html
59. For an example, see the tenor and stylistics employed by Tushar Shah while reviewing an edited volume by IDPAD researchers, published in the year 1990, in *Economic and Political Weekly*. Shah, Tushar (1990) 'Unconvincing Critique of Operation Flood,' *Economic and Political Weekly*, Vol XXV, No 34 (May 26).
60. Parthsarathy, G (1991) 'White Revolution, Dairy Cooperatives and Weaker Sections,' *Economic and Political Weekly*, Vol 26, No 52, (December 28) pp. A177–A183.
61. Ghose, Sanjay (1994) 'It's a Cow's Life,' *Down To Earth*, (April 15) http://www.downtoearth.org.in/node/29681
62. 'The Milk that ate Grass,' *Down To Earth*, (April 15, 1999) http://www.downtoearth.org.in/node/19660
63. 'Importing Cattle Growth Hormone hits snags,' *Down To Earth*, (November 30) http://www.downtoearth.org.in/node/30414
64. Patra, Bijoy Basant (1994) 'Milking the Cattle,' *Down To Earth*, (May 31) http://www.downtoearth.org.in/node/31995
65. 'DDT Milk,' *Down To Earth*, (April 15).
66. 'Treasures in Grass,' *Down To Earth*, (April 15, 1999) http://www.downtoearth.org.in/node/19663 The dire state of affairs on grassland and rangeland as described in this cover story in *Down To Earth* had almost bordered on alarmist proportion. It had severely indicted the government for being "devoid of any policy on grazing or management and protection of the invaluable grasslands". The report reminded grassland scientists that "the last time India's grasslands were surveyed was more than 30 years ago" and quoted B K Trivedi, a scientist at the Indian Grassland and Fodder Research Institute, Jhansi, Madhya Pradesh stating, "If one were to go back (to the areas that were surveyed earlier), the chances of finding rangelands (grasslands) are extremely remote"!
67. 'What's eating the Grasslands,' *Down To Earth*, (April 15, 1999) http://www.downtoearth.org.in/node/19665
68. See, Khurana, Indira (1997) 'Vanishing Breeds', *Down To Earth*, September 15, 1997.

69. See, Mazoomdar, Jay (2013) 'Desi Cow, Almost Extinct,' *Tehelka*, February 2013.
70. Sood, Jyotika (2014) 'New Milky Way,' *Down To Earth*, 15th February 2014, http://www.downtoearth.org.in/content/new-milky-way
71. Patel, Sujata (1990) 'Anand Model' in Doornboss, Martin and K. N. Nair (eds) *Resources, Institutions and Strategies: Dairy Aid and Operation Flood*, Sage Books, New Delhi.

Chapter 9
After the Operation Flood: Studying Transformations in Bovine Holding Pattern

With the advent of neoliberal economic reforms, the private trade interests have entered the dairying business and the share of milk cooperatives in the urban milk market is going down. Two livestock economists have pointed out that currently more than eighty percent of the milk produced in the country is marketed by unorganized sectors and less than twenty percent by the organized sector.[1] The organized sector here involves government and Anand pattern cooperatives, whereas the unorganized sector involves private organizations. As noted in the previous chapter, following the liberalization policies, private dairy trade has also been making headlines, leaving behind the traditionally well-established command of NDDB undermined. There is now an acknowledgement that the cooperatives and the private sector dairy plants today have more or less equal capacities.[2] However, these researchers also put forward a claim that much of the processing capacity by the private sector remains idle, and only about 60% is operated on the day-to-day basis. However, similarly in the government sector, too, most of the primary processing facilities installed in the rural areas are not functional, whereas in the cooperative sector, all the dairy plants are used to their full capacity and remain under-utilized only during the lean period season.

It has been observed that in recent past, the demand for milk and milk products in India has increased rapidly along with achievement of high economic growth.[3] A conjecture is voiced that, chances are that demand growth may surpass output growth. While seeking to analyse the current challenges confronting the Indian dairy sector, these researchers present a conclusion and very strongly articulated policy advice that a multifaceted programme of the nature and extent of Operation Flood is needed.[4]

Thus, the crop-cattle interrelations and transformations under the intensification strategies followed during 'green' and 'white' revolution that we discussed in the foregoing chapters leave us with a question: Has there been any revision in the dominant modes of thinking in such a post-revolutions regime when sustainability concerns have entered the discussion and pastoral perspectives are also beginning to

H. Upadhyaya, *Critical Insights on Colonial Modes of Seeing Cattle in India (1850–1980)*, Asia in Transition 27, https://doi.org/10.1007/978-981-97-1560-2_9

find their place in development discourses? If we look at the work that NDDB and Food and Agriculture Organization are doing within a collaborative programme, we come to understand that NDDB has now decided to assist the small stock-rearing marginalized people. The programme that NDDB and FAO now run is called, South Asia Pro-Poor Livestock Policy Programme, which has a clear agenda to focus on non-bovine stock within livestock-based livelihoods. Recently, researchers have also started to talk about organic animal husbandry in India alongside writings on organic cultivation and food production.[5]

In conversations with key informants at the study site, it was shared that a few rich farmers have sought to adopt milk as a cash crop and dairy farming as an exclusive activity, by focusing only on fodder crop within their cropping cycles. It was noticed that such farmers keep more than 20 cows or buffaloes and supply to village cooperative the milk in bulk. Agricultural journalist, Harish Damodaran also documents some of these farmers who now sell only one crop, viz. milk. These Gen-Next farmers, who are members of KDCMPU now hire workers to manage byres, where one can see milking machines and the milk is pumped directly into an overhead chilling tank in their premises, from where the dairy plant tankers collect it.[6] Three years later, when this theme was revisited by Damodaran, it was found that there are 4293 farmer members of KDCMPU who own at least 30 milch cows or buffaloes each and sell 100 L and more daily.[7] Thus, herd size is increasing and the milk production at village level is getting sophisticated with employment of milking machines, etc. If the primary members have started to reduce the labour on the milking of cows by option for milking machines, at the level of societies, there is now higher number of bulk milk coolers.[8] There are now farmers who tend to look at the milk production as their primary source of livelihoods, and even if they have cultivated land, they have shifted in last one decade to grow only fodder crops. Few farmers that I met in Kheda now devote their entire acreage for napier-hybrid bajra, African tall maize and oats as green fodder. These farmers now try to go in for mechanization that saves them labour time and costs. Thus, recently Gujarat Cooperative Milk Marketing Federation (GCMMF) decided to encourage farmers to buy four bucket milking machines (which costs Rs. 1–1.1 lakh) at a subsidized price of Rs. 60,000. This capacity to invest into mechanization would not have come but for the steadily increasing procurement prices that cooperative members have been getting from GCMMF societies. On the other hand, there are also organic farmers in Kheda who try and keep the crop-cattle interrelations alive. These farmers also explore the marketing channels of their products in niche urban areas through farmers' market route. Civil society activists and academicians concerned about the small-scale, resource poor dairy farmers have also started to voice their concerns that if GCMMF does not check the growing number of large-scale milk producers who own *tabelas* keeping more than 50–60 cows and buffaloes, it might spell a doom for small-farmer dairies.[9]

Even as the country was imagining the revolution in crop cultivation to be followed by another in milk production, Common Property Resources (CPRs) had witnessed a steady decline and increased privatization. N. S. Jodha had pointed out that between the 1950s and 1980s, the area under 'permanent pastures and grazing land' decreased from 70 million hectares in 1947 to 38 million hectares in 1997. Unlike rich farmers with larger operational land holdings, the poor and nomadic livestock owners have greater dependence on the CPRs. While on the one hand, the poor livestock owners' access to grazing land was decreasing, because of the gradual shrinking of commons, the soil and water conservation programmes, so enthusiastically embraced by planners and at times by NGOs also added to their woes by putting those grazing land under 'restrictions' or 'complete ban'.

Since the 1980s scholars have tried to estimate the magnitude of commons and map communities dependent upon them, it was found that states with advanced agricultural systems accounted for 5–7% land in this category (although land devoted to 'cultivated fodder' in the same states had shown impressive double-digit growth!), while those states supporting rain-fed agriculture or pastoralism as main livelihood showed up 25–32% land under commons. The only national level survey to estimate this category, namely National Sample Survey 54th round conducted in 1998, concluded that commons account for 15% geographical area on a national average and amounted to about 0.31 hectares per household. As per this survey, 20% rural households rely on commons for grazing and 25–30% rural households depend upon commons for access to water and almost 45% rural households are dependent on commons for firewood collection.

Today, around three decades after those academic writings that raised concerns on the consequences of crossbreeding on the availability of work animals, the estimated share of plough and cart animals in total power deployed on Indian farms has fallen from 53% in 1971–72 to below 9%. NSS Report No 408 (1991–92) and Report No 493 (2003) found that the trend of declining population of working animals had further accentuated during the decades of 80s and 90s. Despite the onset of Operation Flood, the number of in-milk cattle had declined from 27 million in 1971–'72 to 19 million in 1981–'82, giving support to the argument in this book that in the initial years of Operation Flood , it was an intervention which concentrated more on milk processing, distribution and marketing and not so much on milch animals and their health. The following decade of the 1980s had shown an increase in the proportion as well as absolute number of in-milk cattle in the total cattle population, showing the resurgence with the number of in-milk cattle bouncing back to 30 million in 1991–'92. That trend has however not kept the momentum in the decade of 1990s (in liberalization era) and the absolute number of in-milk cattle has come down to 28 million in 2003. While one needs to be alert to margin of error creeping in due to seasonality (NSSO enumerates data for its round during eight months), it is safe to conclude that the decade of 1980s, under the Operation Flood policies and advocacy of crossbred cattle, would have made resource-rich dairy farmers to increase crossbred cattle in household level herd (microcosm of what NDDB planned as 'national milch herd'). Contrary to this trend of higher contribution from crossbred cattle in the 1980s, during the decade of 1990s, dairy farming seems to have been the era of Murrah and Banni buffalo being touted as black gold and higher level of buffaloization of national milch heard.

This figures of more in-milk buffaloes in livestock census on year-to-year basis needs also to be read along with what is happening to male progeny to buffalo. When

Table 9.1 Different types of cattle as estimated by different NSS reports

	(Numbers in millions) (rural)			
Type of cattle	1971–'72 (26th)	1981–'82 (37th)	1991–'92 (48th)	2002–'03 (59th)
Working	73	66	68	51
In-milk	27	19	30	28
Others	69	76	71	75
Total	169	161	169	154

Source Various NSS Reports

the crossbreeding had come under criticism from social scientists and economists, an argument that was heard was that the Indian peasant does not find a male progeny of buffalo or a male offspring of crossbred cattle beneficial to his agrarian operations. A cursory look at the livestock census figures from Gujarat shows that male buffaloes are not being starved to death and they do reach maturity (3 Years of age) and this probably is linked to the fact that licensed slaughter houses in Gujarat have been reporting higher rates of male buffalo slaughtering (Surat district for example).

During the decade of 90s, the decline of working cattle population was 25%, or 2.5% per year. In contrast to this, the buffalo population in rural India continued to grow from 69 million in 1991–'92 to 76 million in 2002–'03.

Agricultural journalist, Harish Damodaran calls this trend 'Buffaloization of the Nation'. He further notes that percentages of buffalo in bovine stock were higher than national for Haryana (79.3), Punjab (74), Uttar Pradesh (55.8), Andhra Pradesh (54.2), Gujarat (52.4), Rajasthan (47.8) and Bihar (34.8).[10] Table 9.2 represents changes in the last 25 years.

Although the 1982 Livestock Census does not provide figures for Punjab, which was in the midst of political turmoil, the crossbred cattle reported in subsequent censuses shows that proportion of crossbred cattle would have been higher in Punjab, just like the other Green Revolution state, Haryana. In Table 9.1, we have given the major states where share of crossbred cattle is much high. With the only exception of Uttar Pradesh, the number of crossbred amongst cattle has been showing the rising trend and even for Uttar Pradesh, the lack of growth could be explained by that state having switched to milch buffaloes in much large number as preferred milch animal. Ironically, amidst all the talk about 'cow protection' in political discourse, it is in districts of UP such as Mathura, where the number of buffalo figures is quite high and preference for cows is showing a decline as shown in Tables 9.2 and 9.3.

Let us also try and capture the trend in change of bovine stock holding patterns in India for rural and urban area for the decade 1992–2003.

What emerges interestingly from the above data is that even as crossbred cattle registered a spectacular increase during the decade of 1992–2003 (towards end of Operation Flood III phase and liberalization era), buffalo was once again emerging as preferred milch animal for urban India. While for rural India, buffalo population grew by just 15.03%, for urban the growth rate was more than double of that, viz. 39.66%. This means that the model of buffalo-driven urban milch stables was making

Table 9.2 Percentage of crossbred cattle and percentage of buffaloes (1982 and 2007 Livestock Census)

State	Percent of crossbred cattle		Percent of buffaloes in bovine
	1982 census	2007 census	2007 census
Andhra Pradesh	1.31	16.90	54.18
Bihar	0.93	13.34	34.75
Gujarat	0.76	14.32	52.38
Haryana	11.36	36.47	79.31
Himachal	5.70	35.07	25.12
J & K	7.01	47.55	02.33
Karnataka	4.76	20.88	29.17
Kerala	46.92	93.16	03.23
Maharashtra	3.04	19.29	27.29
Nagaland	20.53	54.06	06.93
Orissa	1.76	13.83	08.81
Punjab	N.A	71.94	74.01
Rajasthan	00.29	07.21	47.78
Sikkim	19.08	54.10	00.17
Tamil Nadu	8.54	65.98	15.22
Uttar Pradesh	12.43	10.30	55.77
West Bengal	3.54	13.76	03.82
Delhi	7.69	48.38	75.16
Pondicherry	26.88	92.79	03.75
Total	4.65	16.60	34.60

Source Livestock Census figures from 1982 and 2007

Table 9.3 Trend in change of bovine stock holding in urban and rural India (1992–2003)

Year	Category	Rural (in thousand)	Urban (in thousand)	Total (in thousand)
1992	Crossbred cows	13,462	1753	15,215
	Indigenous	182,425	6944	189,369
	Buffaloes	79,915	4291	84,206
2003	Crossbred cows	21,937	2750	24,686
	Indigenous	153,714	6780	160,495
	Buffaloes	91,930	5993	97,992
Change	Crossbred	62.95% increase	56.87% increase	62.25% increase
	Indigenous	15.74% decline	2.36% decline	15.25% decline
	Buffaloes	15.03% increase	39.66% increase	16.29% increase

Source Dairy India 2007

Table 9.4 Change in cattle population in Kheda during of I and II (1972–1987)

Description	Livestock census (1972)	Livestock census (1987)	Percentage increase or decrease compared to 1972
Cattle (males over 3 years)			
Used for breeding only	614	4242	590.87
Used for breeding and work both	1619	3575	120.81
Used for work only	167,975	120,169	– 28.46
Others	650	975	50
Total male cattle	170,858	128,961	– 24.52
Cattle (females over 3 years)			
In-milk	17,167	49,276	187.03
Dry	17,094	21,687	27.04
Not calved even once	1811	3286	81.44
Working	421	NA	NA
Others	189	743	NA
Total female cattle	36,682	74,992	104.43
Young stock cattle	40,002	67,549	68.86
Total cattle	247,542	271,502	9.67

Source Livestock Census, 1972 and 1987

a comeback. It is also curious that crossbred cattle registered a growth which was four times the growth rate noticed in indigenous cattle. How did this story unfold at the level of Kheda and Gujarat? In Table 9.4, we present the figures for Kheda from the Livestock Census 1972 and 1987 to show the 'Operation Flood' years impact on bovine stock (Table 9.5).

The number of draught bullock population has shown a 24.52% decline in Kheda district during the 15-year period, even as the population of cows showed a spectacular growth by 104.43%. Within that, the growth of in-milk cows had witnessed a magnificent 187.03% growth, driving the constantly rising milk production during Operation Flood years. How has the buffalo population changed in Kheda district during Operation Flood years (Table 9.5)?

Here, we notice that numbers for both male as well as female buffaloes have shown rising trend over the course of fifteen years. Within adult male buffalo across the different use they are put to, the category of those used for breeding and for breeding as well as work have shown growth, while those which are used only for work have shown a decline by 42.07%. Within adult female buffalo, the in-milk category has witnessed an impressive growth by 49.44%. This proves that farmers in Kheda now looked at their bovine stock more from the point of view of its contribution to milk rather than providing draught power.

Table 9.5 Change in buffalo population in Kheda during of I and II (1972–1987)

Description	Livestock census (1972)	Livestock census (1987)	Percentage increase or decrease compared to 1972
Buffalo (males over 3 years)			
Used for breeding only	678	1805	166.22
Used for breeding and work both	123	179	45.52
Used for work only	1754	1016	– 42.07
Others	372	383	2.95
Total adult male buffalo	2927	3383	15.57
Buffalo (females over 3 years)			
In-milk	173,361	259,072	49.44
Dry	110,285	85,725	– 22.26
Not calved even once	13,628	15,057	10.48
Working	986	NA	NA
Others	222	2546	NA
Total adult female buffalo	298,482	362,400	21.41
Young stock buffalo	253,809	282,443	11.28
Total buffalo	555,218	648,226	16.75

Livestock Census 1972 and 1987

Let us try to understand livestock census figures for Gujarat for the period 1997–2012 and understand how the numbers of crossbred cattle is rising. From the data it emerges that in the year 1997, the crossbred cattle constituted 5.08% of total cattle, while in the year 2003, it constituted 8.60% of total cattle. Thus, the preference for crossbred cattle was rising at a spectacular rate compared to indigenous cattle. Male–female ratio in crossbred cattle is noticed at 1:5, while for indigenous cattle, male–female ratio was almost 1:1. Thus, what has been true about the preference for female calf in the case of buffalo was also seen in the case of crossbred cattle. What is also curious is that the livestock census shows the distribution of adult male crossbred cattle within categories as: 'used for work only', 'used for breeding only', 'used for breeding and work both' and 'others'. This leads one to believe that farmers are employing crossbred male cattle as draught and plough bullocks as well, which posits a contrast to what we just discussed above in the case of Kheda during the fifteen-year period of 1972–1987 as shown in Table 9.6.

In the next table, we can witness what was the growth rate of buffalo population in Gujarat during 1997–2003.

What also emerges from Table 9.7 is that during the six-year period (1997–2003), the numbers of crossbred cattle were showing a huge increase—86.83%—whereas

Table 9.6 Number of crossbred and indigenous cattle in Gujarat (1997–2003)

Classification		Year		Percent increase or decrease in 2003 over 1997
		1997	2003	
Crossbred cattle				
Male	Under 1 year	16,000	34,200	113.75
	1–2.5 years	14,000	27,186	94.18
	Over 2.4 years	32,000	56,434	76.35
	Total male	62,000	117,820	90.03
Female	Under 1 year	49,000	93,359	90.52
	1–2.5 years	47,000	88,058	87.35
	Over 2.5 years			
	(a) In-milk as on 15–10-1997/2003	135,000	250,787	85.76
	(b) Dry	35,000	67,108	91.73
	(c) Not calved even once (with other)	14,000	21,833	55.95
	(d) Total (a to c)	184,000	339,728	84.63
	Total female	280,000	521,145	86.12
Total crossbred cattle		342,000	638,965	86.83
Indigenous cattle				
Male	Under 1 year	259,000	301,258	16.31
	1–3 years	315,000	380,457	20.78
	Over 3 years	2,581,000	2,604,368	0.90
	Total male	3,155,000	3,286,083	4.15
Female	Under 1 year	505,000	570,676	13.00
	1 to 3 years	499,000	585,999	17.43
	Over three year			
	(a) In-milk as on 15–10-1997/2003	1,346,000	1,405,897	4.45
	(b) Dry	654,000	713,589	9.11
	(c) Not calved even once (with others)	248,000	222,311	– 10.35
	(d) Total (a to c)	2,248,000	2,341,797	4.17
	Total female	3,252,000	3,498,472	7.57
Total indigenous cattle		6,406,000	6,784,555	5.90
Total cattle (Crossbred + Indigenous)		6,749,000	7,423,520	9.99

Source Livestock Census 1997 and 2003 (*Note* Data for 1997 came in the format with figures rounded to nearest 1000)

Table 9.7 Number of buffaloes in Gujarat (1997–2003)

Classification		Year		Percent increase or decrease in 2003 over 1997
		1997	2003	
Male buffaloes	Under 1 year	283,000	333,673	17.90
	1–2.5 years	159,000	200,741	26.25
	Over 2.5 years	104,000	132,405	27.31
	Total male	544,000	666,819	22.57
Female buffaloes	Under 1 year	922,000	1,089,584	18.17
	1–2.5 years	884,000	1,151,730	30.28
	Over 2.5 years			
	(a) In-milk as on 15–10-2003/7	2,547,000	2,793,530	9.67
	(b) Dry	986,000	1,143,732	15.99
	(c) Not calved even once (with others)	402,000	294,828	– 26.65
	Total (a–c)	3,935,000	4,232,090	7.54
	Total female	5,740,000	6,473,404	12.77
Total buffaloes		6,285,000	7,140,223	13.60

Source Livestock Census 1997 and 2003

the numbers of indigenous cattle were showing an increase by 5.90% and numbers of buffaloes showed an increase by 13.60%.

Now let us see whether the proportion of crossbred cattle within the total cattle population in Gujarat has kept the pace of increasing trend during the decade of 2003–2012 or not. In the year 2007, crossbred cattle constituted 14.32% of all the cattle in Gujarat, whereas in the year 2012, it constituted 19.29 of all the cattle (Table 9.8).

What we notice in the decade of 1997–2007 when we look into the growth rate of indigenous cattle that while during the first five years of the decade the numbers grew by around 6%, in the latter five years numbers grew at less than 1%. What could be the plausible explanation for such a trend? Was it the case—as argued by critics of Operation Flood in early 80s—that the twin trend of rapidly growing milk economy and mechanization of agriculture was making farmers give up on the plough bullocks and practice dairy farming in a fashion where inputs were purchased and crop-livestock integration did not obtain at farm level? Also, how did buffaloization trend perform in the decade of 1997–2007? Here are the figures for buffaloes in Gujarat for the period 2003–2007 showing that in-milk buffalo figures were nearly twice the number of in-milk cows (Table 9.9).

In Table 9.10 are numbers as reported in Livestock Census 2007 and 2012.

Here are the numbers on buffalo from livestock census (2007 and 2012) (Table 9.11).

Table 9.8 Number of crossbred and indigenous cattle in Gujarat (2003–2007)

Classification		Year		Percent increase or decrease in 2007 over 2003
		2003	2007	
Crossbred cattle				
Male	Under 1 year	34,200	69,424	102.99
	1 to 2.5 year	27,186	50,263	84.89
	Over 2.4 year	56,434	91,798	62.66
	Total Male	117,820	211,485	79.50
Female	Under 1 year	93,359	177,515	90.14
	1–2.5 years	88,058	180,669	105.71
	Over 2.5 years			
	(e) In-milk as on 15–10-2003/7	250,787	397,177	58.37
	(f) Dry	67,108	127,799	90.44
	(g) Not calved even once (with other)	21,833	47,544	117.76
	(h) Total (e–g)	339,728	572,520	68.52
	Total female	521,145	930,704	78.59
Total crossbred cattle		638,965	1,142,189	78.76
Indigenous cattle				
Male	Under 1 year	301,258	362,006	20.16
	1–3 years	380,457	422,556	11.07
	Over 3 years	2,604,368	2,531,283	− 2.81
	Total male	3,286,083	3,315,845	0.91
Female	Under 1 year	570,676	632,155	10.77
	1–3 years	585,999	632,402	7.92
	Over 3 years			
	(e) In-milk as on 15–10-2003/7	1,405,897	1,334,414	− 5.08
	(f) Dry	713,589	669,140	− 6.23
	(g) Not calved even once (with others)	222,311	249,579	12.27
	(h) Total (e–g)	2,341,797	2,253,133	− 3.79
	Total female	3,498,472	3,517,690	0.55
Total indigenous cattle		6,784,555	6,833,535	0.72
Total cattle (Crossbred + Indigenous)		7,423,520	7,975,724	7.44

Source Livestock Census 2003 and 2007

Table 9.9 Number of buffaloes in Gujarat (2003–2007)

Classification		Year		Percent increase or decrease in 2007 over 2003
		2003	2007	
Male buffaloes	Under 1 year	333,673	544,933	63.31
	1–2.5 years	200,741	301,926	50.41
	Over 2.5 years	132,405	169,272	27.84
	Total male	666,819	1,016,131	52.38
Female buffaloes	Under 1 year	1,089,584	1,363,308	25.12
	1–2.5 years	1,151,730	1,564,758	35.86
	Over 2.5 years			
	(a) In-milk as on 15-10-2003/7	2,793,530	3,040,460	8.84
	(b) Dry	1,143,732	1,349,517	17.99
	(c) Not calved even once (with others)	294,828	439,395	49.03
	Total (d–f)	4,232,090	4,829,372	14.11
	Total female	6,473,404	7,757,438	19.84
Total buffaloes		7,140,223	8,773,569	22.88

Source Livestock Census 2003 and 2007

How has the Kheda district performed in terms of crossbred cattle and buffaloization? Here in Table 9.12 is data from 1987 to 2012 livestock censuses (note that the figures also include number of young stock within this category). We notice that buffalo female number has always been higher, and while during the period 1850–1900, as discussed in Chap. 3, there had been a 1:3 ratio between cow and female buffalo, and in the year 1972, the ratio between in-milk cattle and in-milk buffalo turned out to be 1:10; now in the period 1987–2012, in all the years the ratio of cow and female buffalo has stayed at 1:6!

Let us also see the change in buffalo numbers for Kheda for the same period (1987–2012) in Tables 9.13 and 9.14. What we can clearly see is that buffalo plays a pivotal role in Kheda's economy, in a much obvious way than crossbred cows.

For Kheda and Anand districts, the number of productive cows (crossbreed), productive cows (Indigenous) and productive buffaloes are as follows according to Integrated Sample Survey 2013–14, published by Department of Animal Husbandry. This data also proves that the contribution of buffalo to the milk revolution story in Kheda is much more pronounced (Table 9.15).

When one looks at the phenomenal contribution of Gujarat in milk production figures, one shall also look into neighbouring districts that witnessed spread out effect of Kheda as a dairying tract very early on. So, as a recent NDDB publication 'Dairying in Gujarat: A Statistical Profile 2013' states on the relative performance of districts: "Banaskantha has the highest number of in-milk buffaloes and cows, with Sabarkantha and Mehsana in second and third position. Sabarkantha has the highest

Table 9.10 Number of crossbred and indigenous cattle in Gujarat (2007–2012)

Classification		Year		Percent increase or decrease in 2012 over 2007
		2007	2012	
Crossbred cattle				
Male	Under 1 year	69,424	Age group has changed	
	1–2.5 years	50,263		
	Over 2.4 years	91,798		
	Total male	211,485	192,542	– 8.96
Female	Under 1 year	177,515	323,129	82.03
	1–2.5 years	180,669	362,569	100.68
	Over 2.5 years			
	(i) In-milk as on 15–10-2003/7	397,177	732,208	84.35
	(j) Dry	127,799	230,633	80.47
	(k) Not calved even once (with other)	47,544	85,622	80.09
	(l) Total (i–k)	572,520	1,048,463	83.13
	Total female	930,704	1,734,161	86.33
Total crossbred cattle		1,142,189	1,926,703	68.69
Indigenous cattle				
Male	Under 1 year	362,006	Age Group has changed	
	1–3 years	422,556		
	Over 3 years	2,531,283		
	Total male	3,315,845	3,024,757	– 8.78
Female	Under 1 year	632,155	791,051	25.14
	1–3 year	632,402	1,148,960	81.68
	Over 3 year			
	(i) In-milk as on 15-10-2003/7	1,334,414	1,910,247	43.15
	(j) Dry	669,140	867,115	29.59
	(k) Not calved even once (with others)	249,579	315,120	26.26
	(l) Total (i–k)	2,253,133	3,092,482	37.25
	Total female	3,517,690	5,032,493	43.06
Total indigenous cattle		6,833,535	8,057,250	17.91
Total cattle (Crossbred + Indigenous)		7,975,724	9,983,953	25.18

Source Livestock Census 2007 and 2012

Table 9.11 Number of buffaloes in Gujarat (2007–2012)

Classification		Year		Percent increase or decrease in 2012 over 2007
		2007	2012	
Male buffaloes	Under 1 year	544,933	Age group has changed	
	1–2.5 years	301,926		
	Over 2.5 years	169,272		
	Total male	1,016,131	835,775	– 17.75
Female buffaloes	Under 1 year	1,363,308	1,950,283	43.06
	1–2.5 years	1,564,758	1,953,765	24.86
	Over 2.5 years			
	(a) In-milk as on 15-10-2003/7	3,040,460	3,534,030	16.23
	(b) Dry	1,349,517	1,544,788	14.47
	(c) Not calved even once (with others)	439,395	566,933	29.03
	Total (a–c)	4,829,372	5,645,751	16.90
	Total female	7,757,438	9,549,799	23.11
Total buffaloes		8,773,569	10,385,574	18.37

Source Livestock Census 2007 and 2012

number of in-milk crossbreds and Kachchh, the highest number of in-milk indigenous cattle. In-milk indigenous cattle like Gir are predominantly spread across Saurashtra region covering Rajkot, Junagadh and Bhavnagar districts; whereas Kankrej are found mostly in northern Gujarat and Kachchh region".

NSS Report No 424, based upon the 50th round (July 1993–June 1994), had shown that the ownership of milch animals and draught animals increases rapidly as one moved from lower size classes of land holding to higher size classes. Thus, while the overall impression that livestock holding shows up more egalitarian distribution patterns compared to landholding, might appear true on a surface level, when it comes to the size of herd, the quality of animals possessed, issues about access to grazing and capabilities to go for cultivated or purchased fodder; the landless, small and marginal land holders have to wage a continuous struggle to survive. NSS Report No 493 (2003: 17) also corroborates this by stating, "the larger size holdings, endowed with greater resources for fodder supply, investments and human labour time, could naturally maintain a larger bovine stock". NSS report pointed out that the differentials in average stock of in-milk bovines between successive categories of holdings had become more pronounced during the 1990s according to the estimates for the 48th and 59th round. In other words, a strengthening of the positive association between in-milk bovine stock and size of household operational holding appears to have occurred in the decade prior to 2002–03. The report also notes that "In the marginal category, the number of in-milk cattle per 100 households was more than the number of in-milk buffaloes through the two decades. The medium and large holdings, on the

Table 9.12 Change in cattle population in Kheda district (1987–2012)

Classification		Year					
		1987	1992	1997*	2003	2007	2012
Crossbred cattle							
Male	Under 1 year	1179	1971	733	2582	2951	5742^
	1–2.5 years	903	1897	816	2737	2678	
	Over 2.4 years	2924	3220	612	6618	9005	2406#
	Total male	5006	7088	2161	11,937	14,634	8148
Female	Under 1 year	5963	7748	2230	4699	11,999	19,571
	1–2.5 years	5818	8422	1457	6037	12,353	22,846
	Over 2.5 years						
	(m) In-milk	17,913	26,636	6092	16,725	29,296	50,868
	(n) Dry	2792	6492	1682	5282	9849	14,814
	(o) Not calved (with other)	841	1334	771	1357	2358	4518
	(p) Total (m–o)	21,546	34,462	8545	23,364	41,503	70,200
	Total female	33,327	50,632	12,232	34,100	65,855	112,617
Total crossbred cattle		38,333	57,720	14,393	46,037	80,489	120,765
Indigenous cattle							
Male	Under 1 year	7742	9681	8275	7705	8441	15666‘
	1–3 years	21,700	21,472	11,692	12,526	14,677	
	Over 3 years	126,037	111,772	80,323	65,224	53,188	60,381 ~
	Total male	155,479	142,925	100,290	85,455	76,307	76,047
Female	Under 1 year	10,545	14,440	11,487	10,016	12,141	15,879
	1–3 years	13,699	15,843	10,589	11,749	12,331	18,741
	Over 3 year						
	(m) In-milk	31,363	46,060	31,040	27,139	28,724	41,167
	(n) Dry	18,895	18,487	11,680	14,814	13,486	18,914
	(o) Not calved (with others)	3188	4056	5826	5701	3556	8999
	(p) Total (m–o)	53,446	68,603	48,546	47,654	45,766	69,080
	Total female	77,690	98,886	70,622	69,419	70,238	103,700

(continued)

Table 9.12 (continued)

Classification	Year					
	1987	1992	1997*	2003	2007	2012
Total indigenous cattle	233,169	241,811	170,912	154,874	146,544	179,747
Total cattle (Crossbred + Indigenous)	271,502	299,531	185,305	200,911	227,033	300,512

Source Livestock Census (* On 2 October 1997, Anand was carved out of Kheda as a separate district, effecting a drop in the numbers in the census thereafter. ^ refers to less than 1.5 years, # refers to over 1.5 years, ' refers to less than 2 years, ~ refers to more than 2 years)

Table 9.13 Change in buffalo population in Kheda district (1987–2012)

Classification		Year					
		1987	1992	1997	2003	2007	2012
Male buffaloes	Under 1 year	21,633	24,049	20,507	19,885	36,858	41143^
	1–2.5 years	7122	10,844	10,728	10,330	18,312	
	Over 2.5 years	3383	4352	1810	3390	6513	6457#
	Total Male	32,183	39,245	33,045	33,614	61,683	47,600
Female buffaloes	Under 1 year	116,375	100,256	69,280	71,277	95,644	122,135
	1–2.5 years	137,313	114,780	53,334	95,440	137,456	177,458
	Over 2.5 years						
	(a) In-milk	259,072	281,437	176,182	186,800	201,827	254,892
	(b) Dry	85,725	98,881	57,578	80,850	99,187	111,616
	(c) Not calved (and others)	17,603	18,911	24,854	20,454	32,026	44,105
	Total (abc)	362,400	399,229	258,614	288,104	333,040	410,613
	Total female	616,088	614,265	381,228	454,821	566,140	710,206
Total buffaloes		648,226	653,510	414,273	488,435	627,823	757,806

Source Livestock Census, several volumes. (^ refers to less than 2 years, # refers to more than 2 years)

other hand, have shown a clear preference for buffaloes in dairy farming, with their average stock of in-milk buffaloes exceeding their average stock of in-milk cattle" (Table 9.16).

The message that unambiguously emerges from this data is that the participation rate of landless within the milk revolution has been gradually going down. Next two Tables 9.17 and 9.18 depict the in-milk stock per 100 households for cattle and buffaloes respectively.

The latest round of NSSO survey on land and livestock holding correlation specifies that the positive correlation between larger landholding and larger herd size continues and now this has been entrenched even for the small ruminant ownership as well. It is found that, generally, the households with larger operational landholdings

Table 9.14 Change in bovine population in Anand district (1997–2012)

Classification	Year			
	1997	2003	2007	2012
Crossbred cattle male	2700	6300	7000	5678
Crossbred female in-milk	13,700	16,800	28,900	46,416
Crossbred cattle female	26,500	32,600	60,100	98,331
Total crossbred cattle	29,100	39,000	67,100	104,009
Indigenous cattle male	24,000	20,000	24,000	15,506
Indigenous female in-milk	30,000	35,000	23,000	36,499
Indigenous cattle female	64,000	72,000	56,000	90,794
Total indigenous cattle	88,000	93,000	80,000	107,300
Total cattle (Crossbred + Indigenous)	117,100	132,000	147,100	211,309
Male buffaloes	16,000	19,000	34,000	32,265
In-milk buffaloes	138,000	131,000	153,000	178,619
Female buffaloes	287,000	273,000	373,000	453,305
Total buffaloes	303,000	291,000	407,000	485,570

Source Dairying in Gujarat: A Statistical Profile 2013–14, NDDB, Anand

Table 9.15 Number of productive cows (crossbred and indigenous) and buffaloes

District	Crossbred Cows		Indigenous Cows		Buffaloes	
	In-milk	Dry	In-milk	Dry	In-milk	Dry
Anand	41,900	20,400	34,900	15,500	174,800	80,100
Kheda	42,200	23,000	39,700	22,300	230,900	159,200

Source Integrated Sample Survey, Gujarat, 2013–'14, pp. 28–30

Table 9.16 Number of in-milk bovine per 100 households

Category	1971–'72 (26th)	1981–'82 (37th)	1991–'92 (48th)	2002–'03 (59th)
Nil	16	7	6	1
Marginal	33	28	41	40
Small	64	48	69	63
Semi-medium	92	74	80	84
Medium	142	106	102	126
Large	225	153	130	208
All	54	37	46	36

Source NSS Reports

Table 9.17 Number of in-milk cattle per 100 households

Category	1971–'72 (26th)	1981–'82 (37th)	1991–'92 (48th)	2002–'03 (59th)
Nil	9	3	3	0
Marginal	23	15	25	22
Small	41	26	38	31
Semi-medium	58	40	39	42
Medium	86	54	43	59
Large	140	86	43	111
All	34	20	26	19

Source NSS Reports

Table 9.18 Number of in-milk buffalo per 100 households

Category	Number of in-milk cattle per 100 households			
	1971–'72 (26th)	1981–'82 (37th)	1991–'92 (48th)	2002–'03 (59th)
Nil	7	4	3	0
Marginal	10	13	16	18
Small	23	22	31	32
Semi-medium	35	34	41	42
Medium	56	51	59	67
Large	85	67	87	97
All	20	17	20	17

Source NSS Reports

(i.e. medium or large) have higher number of bovine and sheep and goats compared to the households with smaller operational holdings (e.g. landless, marginal, small or semi-medium), while the situation is reverse in case of ownership of pig and poultry. What is curious is that within the landless, there is a preference for buffaloes compared to cattle, even as this landholding category's possession of in-milk bovine stock is said to be 1 per 100 households, in the last NSSO report in 2002–'03 (Table 9.19).

From this information based on the latest round of NSSO, we can see that in Gujarat on an average, there is a preference towards buffalo stock within bovine livestock holding. This clearly suggests the spread effects of Kheda and proves that milk production is not merely a subsidiary livelihood option for farmers here, but it has emerged as a primary livelihood option for many resource-rich farmers. While discussing the buffalo numbers and growth rate in proportion female buffalo population within bovine stock, we also presented some observations from field that represent the changing nature of milk cooperatives. There is higher diffusion of milking machines in Banaskantha district, facilitated by GCMMF subsidy, and there has come up some sort of rural *tabelas* (milch stables) in this district. In the next chapter, there is an effort to present conclusion and key learnings from this research.

Table 9.19 Number of households owning and average of animals owned per 1000 households

	Number of house reporting owning per 1000 HHs in Gujarat	Average number of animals owned per 1000 households in Gujarat
Cattle (adult male)	183	355.70
Cattle (adult female)	266	412.10
Cattle (adult male + female)	371	767.80
Cattle (young male)	68	87.30
Cattle (young female)	177	272.70
Cattle (young male + female)	217	360.00
Buffalo (adult male)	44	60.70
Buffalo (adult female)	436	730.30
Buffalo (adult male + female)	456	791.10
Buffalo (young male)	59	80.90
Buffalo (young female)	300	456.80
Buffalo (young male + female)	335	537.70

Source Key Indicators of Land and Livestock Holding in India p. A23 and A24

Notes

1. Rajendran, K. and Samarendu Mohanty (2004) 'Dairy Cooperatives and Milk Marketing in India', *Journal of Food Distribution Research*, July 2004, pp. 34–41.
2. Datta, T. N. and B. K. Ganguly (2002) "Analysis of Consumer Expenditure Pattern in States with Special Reference to Milk and Milk Products," National Information Network, National Dairy Development Board. Cited in Rajendran and Mohanty (2004).
3. Gautam; Dalal, R. S. and V. Pathak (2010) 'Indian Dairy Sector: Time to Revisit Operation Flood', *Livestock Science*, Vol 127, No 2–3, pp. 164–175.
4. Their policy advice appears to have had an impact and Natioanl Dairy Development Board sought a World Bank loan to start an ambitious National Dairy Development Plan with a timeline 2010–2017.
5. Verma, A. M.; Singh, P. K. and Pandey, P. (2008) 'Organic Animal Husbandry in India', *Indian Farming*, July, pp. 29–30.
6. Damodaran, Harish (2012) 'Amul's Gen-Next farmers', Buisness Line, 20th November, 2012; http://www.thehindubusinessline.com/opinion/columns/harish-damodaran/amuls-gen next-farmers/article4115993.ece last accessed on 8th June 2016.
7. Damodaran, Harish (2015) 'the Big Picture: A new churning', The Indian Express, 25th January 2015 http://indianexpress.com/article/india/india-others/the-big-picture-a-new-chu rning/99/ last accessed on 10th June 2016.
8. For instance, Damodaran in the newsclip reports that in Banaskantha District Cooperative Milk Producers' Union, there are bulk milk coolers in 1060 societies out of total 1356, a much higher frequency compared to one third of GCMMF societies that report possessing bulk milk coolers.

9. See for instance, Ramdas, Sagari (2015) 'Death of Small-Farmer Dairies amidst India's Dairy Boom', *Economic and Political Weekly*, Vol L, No 19, pp. 21–23.
10. Damodaran, Harish (2012) 'Cow Belt or Buffalo Nation,' *The Hindu Business Line*, 18th April 2012. http://www.thehindubusinessline.com/opinion/columns/harish-damodaran/article3328532.ece

Chapter 10
Conclusion

This study has traced the history of interrelations between bovine stock and crops in Western India. Through archival research, it has engaged with questions that confronted colonial experts working on the issues of cattle breeding, plant breeding, veterinary science and organized dairying. It started by reviewing scholarly literature on the history of pastoral and agro-pastoral society in India, which discusses a fraught symbiosis that existed between crop cultivation and livestock rearing. In part, this recognition of thick set of relationship and interdependence between transhumant pastoralists and settled cultivators gets articulated in the work and writings of veterinary experts working with Royal Veterinary Corps as well as in writings of Sir Albert Howard, who came to India to work at IARI, Pusa, as imperial economic botanist. By paying close attention to Howard's initial work on Indian wheat at IARI and his dissenting opinions on the direction agricultural research was taking in the early 1920s, it has been argued how a discourse on 'mixed farming' as a desirable evolutionary stage of agrarian transformations had emerged. The 'surplus cattle' notion and its discursive influences are discussed alongside tracing the discontent that a blanket endorsement of 'surplus cattle' discourse and its corollary policy implication that sought to increase the slaughter rate generated amongst nationalist leaders who articulated 'cow protection' discourse. By reviewing the contemporary statistical details that speak about the economic contribution by cattle, an effort is made to arrive at a better understanding on scholarly opinions by anthropologists and economists who argued that the cattle population in India had more to do with distinctive features of ecological and economic factors of agrarian and agro-pastoral transformations. By situating the emergence of 'surplus cattle' discourse as well as 'mixed farming' discourse, it underlines the dissenting voices that had repeatedly cautioned against the shift of agrarian and agro-pastoral ecosystem towards accelerated intensification that had a potential to lead to disintegration of crop-cattle-livelihood interrelations.

The colonial interventions in plant genetics and a policy of agricultural research were biased towards political exigencies of empire by giving cash crop, such as cotton promotion preference over the cereals. The narrative on the emergence of

H. Upadhyaya, *Critical Insights on Colonial Modes of Seeing Cattle in India (1850–1980)*, Asia in Transition 27, https://doi.org/10.1007/978-981-97-1560-2_10

provincial and imperial departments of agriculture and veterinary at the turn of the century (1868–1905) shows that a series of famines had forced the late colonial State to initiate some action on agriculture and livestock development front that was not merely confined to militaristic motives. However, when one reviews the discussions amongst colonial experts on agriculture and veterinary science about the cattle breeding policy and objectives, one finds that the colonial policies on cattle breeding revolved around introducing more and more milk strains in *non-de-script* stock. Even as 'dual purpose' cattle breeding policy was repeatedly talked about, there were differences of opinion on whether the milk and draught characteristics were physiologically compatible or not. In this late colonial period, experts did pay some attention to addressing the question of city milk supply and initiated acclimatization studies for a few well-known indigenous breeds such as Sahiwal, Gir and Sindhi.

However, by the late colonial period, the adverse impacts of restrictions on forest grazing and extension of perennial irrigation on indigenous breeds of cattle had started to appear. Similarly, there were the ill-effects arising from an unregulated export of well-known indigenous breeds of milch cattle. While the late colonial state made interventions in the sphere of organized dairying and dairy education—initially with militaristic interest, invoking the response from Kheda peasants in the form of milk boycotts in the 1920s—on the question of crossbreeding with exotic bulls, experts had walked a cautious path. It becomes clear from those early discussions that concerns were expressed to keep crossbreeding experiments restricted to Military Dairy Farms and few plantation sites in hills where climatic conditions were favourable. While discussing and debating on cattle breeding policy, experts in Bombay presidency had also stressed upon keeping the agriculture and livestock interrelations alive.

The role played by urbanization and emerging milk markets in cities and the urban middle-class concerns on the issue of adulteration of milk received favourable responses from colonial administration. It also gleans through rare—and hitherto unexplored—insights from archives about the institutional model of milk cooperatives and brings to the fore arguments fervently made by imperial experts on replication of the Calcutta model of cooperativization of city milk supply. It also discussed the place government research creamery at Anand held for W. Smith and Col. Arthur Olver, who served as imperial dairy expert and imperial expert on animal husbandry, respectively. The harms that the specialized form of dairy farming based on city milch stables had caused to well-known milch breeds have been studied. The colonial efforts to rehabilitate milch animals away from the milch stables located in the heart of the cities have also been reviewed.

An attempt has been made to study factors that led to emergence of Kheda as a premier dairying tract in Bombay presidency. It explains how from very early on (1880–1920) the trend towards buffaloization of bovine stock had taken roots in Kheda district. While in India till 1966 the livestock number showed the ratio of 1:3 for female buffalo in comparison to cows, in Kheda this ratio was 3:1. Did such a pronounced buffaloization create a scarcity of plough bullocks for Kheda patidars? From the contemporary studies that look at peasants in Kheda, it becomes clear that peasants in Kheda were sharing their work animals.[1] The role of the new technological breakthrough in the form of artificial insemination was attracting the attention of experts and the proposal of introducing the same in India was made around late 1930s. It also shows how Kheda had by then become a site where private trade in milk and milk products thrived. It showed that there existed many firms besides Polsons—although Polson seems to have emerged as the largest—and firms like Doshi Brothers also indulged in lobbying efforts with British state in the early 1940s. It also brings to the fore, hitherto unexplored literature from archives that states how a 1939 expert committee report for cattle and dairy development in Bombay had proposed to organize cattle breeding and milk supply cooperative societies in Kheda district. This report had also made a suggestion that government research creamery at Anand be handed over to such cooperative societies union when such an organization comes up. However, milk cooperatives did not take a sustained momentum in Kheda till the year 1946.

In post-colonial period, this research study reviews two major phases of India's development programmes, 'green' and 'white' revolutions and how these unfolded by studying contexts and actors that played pivotal role. In Chap. 5, Gandhian critique of the ideology of productivity is reviewed and arguments put forward by Gandhian thinkers such as J. C. Kumarappa and Satish Chandra Dasgupta are discussed. It also presents a detailed account of factors that shaped up the emergence of food policy in the aftermath of the Bengal Famine and the partition. The narrative brings under focus rapidly changing perspectives of Indian planners, who moved from the community development paradigm to that of intensification. Continuing with the in-depth perusal of discussions at meeting of animal husbandry wing as well as crops and soil wing of ICAR, it presents an analysis of how the post-colonial experts viewed the task of 'settling the nomads' and improving the 'grasslands'. It shows how post-colonial state continued to display a sedentary bias that adversely impacted pastoral people, despite a few voices that expressed dissent to the proposals of locking nomadic communities down to a piece of land and curbing their mobility.

It has been argued that while the early proposals on the question of city milk supply had laid emphasis on cooperativization agenda, there curiously arose a pattern that imagined cooperative organization of milk producers to be restricted to procurement side, while the processing and distribution was designed to be handled by top-heavy bureaucratic organization. It also talks about the emergence of milk colonies around cities on Aarey pattern and under the Food and Agriculture Organization (FAO) assistance. It also presents how the Key Village Scheme performed in the sphere of cattle development and how the 'intensification' agenda led to this being replaced by Intensive Cattle Development Programme (ICDP). It discusses 'green' revolution-induced changes in cropping pattern in Kheda alongside looking at what 'green' revolution paradigm had proposed in terms of restructuring of bovine stock holding pattern. The ideologues of 'green' revolution suggested to introduce mechanization of agriculture and increased use of chemical fertilizer. While planners had hoped that such a shift will lead to farmers investing into increasing proportion of milch animal in their herd, review of contemporary studies suggests that this restructuring process did not unfold equally in all 'green' revolution sites. By looking at the cropping pattern changes as well as bovine stock holding changes in Kheda following 'green' revolution, it has been attempted to present how the nature of cattle and dairy development was changing. During the phase of 1961–1972, the livestock census data suggests that 'Anand pattern' of buffaloization driving dairy development had started to find wider diffusion in other promising milk tracts in Gujarat.

During the five-year period—1961 to 1966—when intensification strategy was around the corner and KDCMPU's growth had attracted the national attention, we witnessed that even as male cattle population registered a sluggish growth—and within this the rapid decline in breeding bull population suggested a rapid diffusion of artificial insemination technology—female cattle population decreased by 3% and female buffalo population registered a considerable growth at 7% in Kheda. In the very next five-year period (1966–1972), this buffaloization of bovine stock trend got even more entrenched in Kheda district, where the 11th Livestock Census (1972) figures showed the number of in-milk buffaloes to be ten times higher than the number of in-milk cattle! Within this period, the number of cattle had shown decline by 12.45%, whereas buffaloes had shown an increase by a whopping 30.34%! We also find this buffaloization trend catching up in three other districts of Gujarat—namely, Sabarkantha, Mahesana and Surat, which showed falling numbers of cattle stock and rising proportion of buffalo within bovine stock holding of farmers. Towards latter part in chapter, a different sort of intensification that was proposed by the then Grassland and Fodder Advisor to Government of India is analysed. By engaging with the vocal critique of cereal-focused planning of Third Five Year plan, this section also points out how R. O. Whyte had advocated to increase acreage under cultivated fodder.

Next, a narrative has been presented on factors that shaped the advocacy for replicating Anand pattern after that public announcement by Lal Bahadur Shastri calling upon a young dairy technocrat to drive the dairy development agenda. It has been argued in this chapter that young Varghese Kurien had hit upon the idea of a national level milk and milk product marketing board for boosting the incomes of milk producers, he had outlined in a blueprint in 1957 itself. Sifting through the dusty pages of archives led this researcher to a published copy of that blueprint which was presented by Asst. General Manager (H. M. Dallaya) in August 1957 to Dr. Burde who worked as Asst. Professor of Dairy Technology and Dairy Administration at NDRI, Karnal, suggesting that KDCMPU had engaged in protracted advocacy for this proposal. From contemporary sources, it has been argued that KDCMPU was emerging as a model worth replicating and there appeared numerous articles and editorials that took that line while comparing it with Aarey.

It also brings narratives that have not found expression in the linear and celebratory history of milk revolution. It recalls other personalities who made significant contributions in the task of dairy development such as J. S. Grewal in Punjab, M. S. Palaniappa Mudaliar in Coimbatore, L. C. Sikka in Calcutta and Dara N. Khurody in Bombay. It also shows that the milk production targets at national level had posed challenge to the planners during the onset of fourth plan. It engages with an analysis of contemporary writings by planners and advisors in the late 1960s to bring out the complexity of the issue of India's efforts to attain self-sufficiency in terms of milk availability. Earlier literature that studies Operation Flood is often found lacking an in-depth understanding of the decade of 60s, when the per capita milk availability figure had remained below the ICMR suggested nutrition standards, even as milk products plants and capacity addition in state sector was increasing. In the latter part of this chapter, the Operation Flood phase is discussed, alongside the review of academic criticism of the programme and the responses from NDDB.

Next, the research study presents the period after Operation Flood by highlighting the buffaloization of bovine economy and the diffusion of crossbred cattle by presenting data from Livestock Census (1987–2012 for Kheda and 1997–2012 for Gujarat), as well as recent NSSO rounds. During this fifteen-year period (1972–1987), number of draught bullock population in Kheda had witnessed a 24.52% decline, even as the population of cows showed a spectacular growth by 104.43%. Within that growth of cows, the growth of in-milk cows had witnessed a magnificent 187.03% growth, driving the constantly rising milk production from Kheda during Operation Flood years. Now, there was also a rapid diffusion of crossbred cows—although proportionately at a lesser rate than buffaloization—that was taking place in Kheda. The adult female buffalo number had witnessed an increase by 21.41%, and within this, the rise of in-milk buffalo category accounted for a spectacular 49.44%.

It also presents an in-depth analysis of the livestock census figures for Gujarat for the period 1997–2012, whereas for Kheda, it presents an analysis of livestock census figures for the period 1987–2012. We notice that for Kheda as well as for Gujarat, the number of crossbred cows and buffalo shows a spectacular rise. While in the case of Kheda, numbers of breed able crossbred cows witnessed a rise from 24,455 to 81,623, numbers of breed able indigenous cows had witnessed a rise from 60,295 to 78,451.[2] Thus, one can safely observe that the proportion of crossbred cows within bovine stock is rising at a higher rate than indigenous cows. The number of breed able buffaloes has grown at a much higher rate, and it accounts for a much higher percentage of bovine stock today.

Compared to the decade of the 1980s, today there are more community-based organizations that work to revive and revitalize indigenous breeds. Even the National Dairy Development Plan (2011–2017) overtly acknowledged that even those indigenous breeds such as Gir and Ongole need to be brought back into discussions. During the course of writing this thesis, this researcher had an opportunity to interact with animal scientists working at National Bureau of Animal Genetic Resources and witnessed intense debates on continued reliance on crossbreeding agenda. Sustainability has emerged as a major concern and policy and programme choices today reflect on resource constraint with a sensitivity displayed that was never seen before. The numbers of exotic/crossbred milch cattle have kept rising. While the 18th livestock census reported a 28.3% increase, the latest livestock figures show their numbers rising by 34.78%. However, whether higher levels of crossbred cattle population can be sustained under the ever-present feed and fodder crisis has emerged as a question that has started to puzzle animal scientists.

One of the key transformations that we have noticed over the last three decades since the onset of 'white' revolution has been the decreasing capacity of poor and landless households to keep in-milk bovines. The historically established pattern of interdependence of livestock ownership on Common Property Resources has given way to a more positive correlation between land ownership and the capacity to keep in-milk bovine livestock. These processes may be seen as integral to the larger transformations brought about by the 'green' and 'white' revolutions. Today however, a more critical understanding of the logic and long-term consequences of these interventions provides us with a vantage of looking at the problem from what we might broadly call a 'post-revolution perspective'. Amidst the calls for the second 'green revolution' and 'operation flood' type intervention, we live in post-revolution times, wherein we need to aspire for nutritional and ecological security in the place of Green Revolution and Operational Flood type 'food security'. We are in an urgent need of building an architecture that revitalizes the symbiotic relationship between crop cultivation and livestock keeping along the sustainability axis. Our development programmes in animal husbandry need to display a clear bias for landless and marginal farmers.

In the study site Kheda, and in Gujarat generally although we find that amongst poor and marginalized people, there is a tendency off late to keep goats and buffaloes, even as in-milk proportion within the buffalo owned by poor, landless persons are negligible. This poses a curious question, whether the landless person's buffalo holding contributes to milk production, or does that preference underline contribution to meat production as well. With the advent of higher economic growth, the size of livestock products has gone up within the dietary choices of people.

While India might have registered a tremendous growth in milk productivity thanks to buffaloization and rapid diffusion of crossbreeding with imported milch cattle breeds, 2007 and 2012 Livestock Census throw up figures that require us to reflect upon the question of feed and fodder requirement to sustain this rally. The recently initiated World Bank-assisted project, National Dairy Plan Phase I, is being implemented in 189 districts out of the 427 districts of 14 states, and livestock holding pattern that these states show is such that "buffaloes constitute nearly half (49%) of the milch animals in project area and the balance is distributed almost equally between Crossbred cows (26%) and Indigenous cows (25%)". Again, within these 14 spearhead states, Haryana, Gujarat, Uttar Pradesh, Punjab, Andhra Pradesh and Rajasthan have significantly higher proportion of buffalo population, ranging from 82% in Haryana to 63% in Rajasthan. Similarly, Kerala and Tamilnadu have respectively 84% and 74% crossbred cattle population. With rapid diffusion of crossbreeding and higher rate of buffaloization, these states have emerged as 'advanced' states in terms of milk production, the gap between the demand and availability of fodder and feed has been steadily increasing. The percentage land under 'cultivated fodder' has remained around 4–5% and planned interventions in this area also run the risk of widening the divide amongst rich and poor farmers. The access to 'cultivated' or 'purchased' fodder (unlike grazing) is almost always mediated by class status.

Sadly, NSSO has not revisited the commons issue after that sole effort in the 54th round in 1998. It is also the need of the hour to examine how the livestock holding patterns have been changing in rain-fed areas and what policy interventions are needed there. While on the one hand, MGNREGA and Forest Rights Act might have opened some space for ensuring sustainable options for nomadic and pastoral communities, there is a need to reorient cattle and milk-oriented livestock development policy towards small ruminants. There is a need to restore symbiosis between agrarian and pastoral concerns, between cultivator and nomad and last but not the least between crop cultivation and cattle rearing.

Notes

1. This phenomenon has often been referred to as "India's ancient Uber Economy". See, Narain, Sunita (2019) "India's Cow Crisis Part 3: Brutal to Kill India's Ancient Uber Economy", *Down To Earth*, (8th January 2019), https://www.downtoearth.org.in/news/agriculture/india-s-cow-crisis-part-3-brutal-to-kill-india-s-ancient-uber-economy-62752
2. We need to remember that the 2012 numbers do not include those areas which were earlier part of Kheda and are now part of a newly carved out Anand district.

References

Primary Documents

Agriculture and Animal Husbandry in India for the years 1933-'34, Imperial Council of Agriculture Research, New Delhi.

Agriculture and Animal husbandry in India 1934-'35 and 1935-'36, Imperial Council of Agriculture Research, New Delhi.

Agriculture and Animal Husbandry in India for the year 1938-'39, Imperial Council of Agriculture Research, New Delhi.

Annual Administrative Report of the Civil Veterinary Department of Bombay Presidency for the year 1894–95, Government of India Press, Shimla.

Annual Administrative Report of the Civil Veterinary Department of India for the year 1899–1900, Government of India Press, Shimla.

Annual report of the Department of Agriculture, Bombay Province, 1940-'41, Government of India Press, Shimla.

Annual Report of the Department of Agriculture, Bombay Province, 1941-'42, Government of India Press, Shimla.

Annual Report of the Department of Agriculture, Bombay Province, 1944-'45, Government of India Press, Shimla.

Annual Administrative Report of the Department of Animal Husbandry and Veterinary Science, Bombay, 1950–51.

Annual Report on Working of Co-operative Societies in Bengal for the year 1944, Government Printing Press, Alipore, Bengal.

Annual Report on Working of Co-operative Societies in Bengal for the year 1945, Government Printing Press, Alipore, Bengal.

Annual Administrative Report of the Department of Animal Husbandry and Veterinary Science, Bombay, 1950-'51, Government of India Press, Shimla.

Bradley, F. S. H. (1911) *The Indigenous Breed of Cattle in Rajputana*, in Breeds of Indian Cattle Series, Civil Veterinary Department, Government of India Press, Shimla.

Department of Agriculture, Punjab (1910) *Cattle and Dairying in Punjab*, Department of Agriculture, Punjab.

Editorial (1872) The Indian Rinderpest, *The Calcutta Review*, Vol 55, No 109, Pp. 169–186.

Editorial (1916) *The Journal of Dairying and Dairy Farming in India*, Vol IV, Part 1, October 1916, Dairy Farming Association of India, New Delhi.

Editorial (1917) *The Journal of Dairying and Dairy Farming in India*, Vol IV, Part 3, Dairy Farming Association of India, New Delhi.

H. Upadhyaya, *Critical Insights on Colonial Modes of Seeing Cattle in India (1850–1980)*, Asia in Transition 27, https://doi.org/10.1007/978-981-97-1560-2

Five Years of Research on Dwarf Wheats, Indian Agricultural Research Institute, New Delhi, INSDOC Press, 1968.
Government of India (1961) *Review of Key Village Schemes*, Ministry of Agriculture, GoI, New Delhi.
Hewlett, K. (1879) *Breeds of Indian Cattle: Bombay Presidency*, Government Printing Press, Calcutta.
Interim Report of the National Commission on Agriculture on Milk Production through Small and Marginal Farmers and Agricultural Labourers, Ministry of Agriculture, Government of India, December 1971.
Miscellaneous Bulletin of Imperial Council of Agricultural Research, No 27, Delhi, Manager of Publication, 1939.
Observations on Indian Agriculture, A Report of United States Technical Cooperation Mission, New Delhi, March 16, 1953.
Patel, M. D. and C. N. Dave (1944) Kankrej Husbandsmen of Gujarat, *Indian Farming*, Vol V, No 5, (July 1944).
Proceedings during Question Hour in Rajya Sabha, 2nd December 1983.
Proceedings of the Third Meeting of Board of Agriculture in India, Superintendent of Government Printing, Calcutta.
Proceedings of the Fourth Meeting of Board of Agriculture in India, Superintendent of Government Printing, Calcutta.
Proceedings of the First Meeting of Veterinary Officers in India, Superintendent of Government Printing, Calcutta.
Proceedings of the Second Meeting of Veterinary Officers in India held at Calcutta (26 February to 2 March 1923), Government Printing Press, Shimla.
Proceedings of the Second Meeting of the Crops and Soil Wing of the Board of Agriculture and Animal Husbandry in India (held at Lahore from 6th to 9th December 1937), Government of India Press, Shimla.
Proceedings of the Third All India Sanitary Conference held at Lucknow in January 1914, Government of India Press, Shimla.
Proceedings of the Bombay Provincial Board of Agriculture held at Poona on 8th and 9th June 1920, Government of India Press, Shimla.
Proceedings of the Second Meeting of Animal Husbandry Wing of Board of Agriculture and Animal Husbandry, Government of India Press, Shimla.
Proceedings of the Fourth All India Food Conference held at New Delhi on 13th, 14th, 15th and 16th October 1943, Government of India Press, Delhi.
Proceedings of the Seventh Meeting of Crops and Soils Wing of Board of Agriculture and Animal Husbandry in India (held at Madras from the 7th to 10th of April 1948), Government of India Press, Shimla.
Report on India's Food Crisis and Steps to Meet It, The Agricultural Production Team (sponsored by The Ford Foundation), Ministry of Food and Agriculture, Government of India, April 1959.
Report of Community Projects and National Extension Service (1957), Vol III, Part 1.
Report of the Indian Famine Commission (1898), Superintendent of Government Printing, Calcutta.
Report of the Third Humanitarian Conference, Bombay held on 2nd and 3rd September 1918, The Bombay Humanitarian League, Bombay.
Report of the Second Indo-American Team on Agricultural Education, Research and Extension, Indian Council of Agricultural Research, July 1960.
Review of Agricultural Operations in India for the years 1929–30 & 1930-'31.
Report of the Committee on Livestock Feeds and Fodder, 1965, Central Council of Gosamvardhan, New Delhi.
Royal Commission on Agriculture, Vol II, Part I (Evidence taken in Bombay Presidency), 1928, Government of India Press, Shimla.
The Collected Works of Mahatma Gandhi, Vol XXVIII (August to November 1925), The Publications Division, Ministry of Information and Broadcasting, April 1968, Pp. 390–391.

The first Annual Administrative Report of the Civil Veterinary Department in India for the official year 1892–93, Government of India Press, Shimla.
The Indian Taxation Enquiry Committee (1925), Vol III, Evidence Delhi and Benares, Central Publication Branch, Government of India, Calcutta.
The Veterinary Record, Vol IV, No 194, (26th March 1892).

Secondary Sources

Adcock, Cassie (2019) Preserving and Improving the Breeds': Cow Protection's Animal Husbandry Connection, *South Asia: Journal of South Asian Studies*, Vol 42, No 6, pp.1141–1155.
Adcock, Cassie & Radhika Govindrajan (2019) Bovine Politics in South Asia: Rethinking Religion, Law and Ethics, *South Asia: Journal of South Asian Studies*, Vol 42, No. 6, Pp. 1095–1107.
Adhiya, Nandlal Chhaganlal (1925) *Vegetable Ghee ane PraNi Raksha* (In Gujarati), retrieved from archives.org.
Agarwal, Shambhoo Nath (1943) *India's Food Problem and Price Control*, Business Bureau, Commerce Department, Allahabad University.
Alvarez, Claude (1985) Imperialism through Food Aid: The Role of Third World Elites, in Alvarez, Claude (ed) *Another Revolution Fails*, Ajanta Publications, New Delhi.
Alvarez, Claude (1987) Body-blow, *Illustrated Weekly of India*, 19th July 1987. http://www.indianet.nl/a870719.html
Amble, V. N. (1965) Feed requirement of bovines and possibilities of meeting them, *Indian Journal of Agricultural Economics*, Vol 20, No 1, (Jan-March), pp. 7–76.
Amble, V. N.; Murthy, V. V.; Sathe, K. V.; and Goel, B. B. P. S. (1965) Milk production of bovines in India and feed availability, *Indian Journal of Veterinary Science and Animal Husbandry*, No 35, pp. 221–233.
Anonymous (1973) Draining the Villages? *Economic and Political Weekly*, January 20, 1973.
Anonymous (1985) Milking the Land, in *State of India's Environment* 1984–85, Centre for Science and Environment, New Delhi, Pp. 15–16.
Anonymous (1992) Importing Cattle Growth Hormone hits snags, *Down To Earth*, (November 30, 1992). https://www.downtoearth.org.in/news/importing-cattle-growth-hormone-hits-snags-30414
Anonymous (1994) DDT Milk, *Down To Earth*, (April 15, 1994).
Anonymous (1997) 65 Million Draught Animals, *Down To Earth*, (October 31). http://www.downtoearth.org.in/node/11947
Anonymous (1999) What's eating the Grasslands, *Down To Earth*, (April 15, 1999). http://www.downtoearth.org.in/node/19665
Anonymous (1999) Treasures in Grass, *Down To Earth*, (April 15, 1999). http://www.downtoearth.org.in/node/19663
Anonymous (1999) The Milk that ate Grass, *Down To Earth*, (April 15, 1999). http://www.downtoearth.org.in/node/19660
Anonymous (1956) Calcutta Milk Scheme under FAO Auspices, in *The Economic Weekly*, September 15, 1956. Pp. 1088–1089.
Baber, Zaheer (2008) Review of *Ecology, Colonialism and Cattle*, in *Agricultural History*, Vol 82, No 1, Pp. 109–110.
Bansil, P. C. (1954) India's Food Resources and Population, Vora & Co Publishers, Bombay.
Bates, Crispin (1981) The Nature of Social Change in Rural Gujarat: The Kheda District, 1818–1918, Modern Asian Studies, Vol XV, No 4, pp. 771–821.
Bhatia, B. M. (1970) *India's Food Problem and Policy since Independence*, Somaiya Publications, Bombay.
Bhattacharya, Anjali (1986) EEC Dairy Aid eroding Self-Reliance Efforts, *Hindustan Times*, 15th March 1986. http://www.indianet.nl/a860315.html

Bhalla, G S and Y K Alagh (1979) *Performance of Indian Agriculture: A District wide Study*, Sterling Publishers, New Delhi.

Bhattacharya, P (1951) Note on 'The Possibilities of Introducing Artificial Insemination on a Field Scale in Suitable Areas', Note prepared for the Eighth Meeting of Animal Husbandry Wing of the Board of Agriculture and Animal Husbandry in India held at Mysore on 17 to 22 February 1949, Government of India Press, Simla, Pp. 160–164.

Biswas, James Nirendra Kumar (1948) Calcutta Cooperative Milk Societies Union, *The Indian Cooperation Review*, Vol XIV, No 03 (July–September).

Bose, Chunilal (1910) Necessary Measures for Prevention of Food Adulteration, Calcutta Medical Journal (July 1910).

Bose, Chunilal (1918) *The Milk Supply of Calcutta,* Central Press, Calcutta.

Bose, J. P. (ed) (1924) *The Scientific and Other Papers of Rai Bahadur Chunilal Bose* (Volume II), Forward Press, Calcutta.

Brown, Dorris (1971) *Agriculture Development in India's Districts*, Harvard University Press, Cambridge, Massachusetts.

Burt, Bryce (1942) Agricultural Progress in India during the decade of 1929–1939, *Journal of Royal Society of Arts*, Vol 90, No 4607, pp. 202–223.

Carruthers, Ian (1978) Raking up Mud and Magic for Agricultural Progress, *Economic and Political Weekly*, Vol 13, No 20, p. 834.

Casimir, Machael and Aparna Rao (2004) *Nomadism in South Asia*, Oxford University Press, New Delhi.

Caton, Brian P. (2013) *Writing the History of a Colonial Institution: The Case of the Government Cattle Farm, Hissar*; NMML Occasional Paper No 38, History and Society (New Series), Nehru Memorial Museum and Library, Teen Murti House, New Delhi.

Caton, Brian P. (2014) The Imperial Ambition of Science and its Discontents: Animal Breeding in Nineteenth Century Punjab, in Rangarajan, Mahesh and K. Sivaramakrishnan (eds) *Shifting Ground: People, Animals and Mobility in India's Environmental History*, Oxford University Press, Delhi, pp. 132–154.

Chakravarthy, A K (1985) Cattle Development Problems and Programs in India: A Regional Analysis, *GeoJournal*, Vol 10, No 1, pp. 21–45.

Chakravarti, A. K. (1987) 'Availability of Cattle Fodder in India,' *Geographical Review*, Vol 77, No 2, pp. 209–217.

Chatterjee, I (1960) Milk Production in India, *Economic Weekly*, (September 03, 1960), pp. 1347–1348.

Chatterjee, I. (1962) A Single Species Cattle Economy, *Agricultural Situation in India*, Vol 17, No 7, Pp. 764–784.

Chatterjee, Nilananda (1926) A Note on the Cattle of India in *The Condition of Cattle in India*, All India Cow Conference Association, Calcutta.

Coatman, J. (1928) *India in 1926-'27: A Statement prepared for Presentation to Parliament in accordance with the requirements of the 26th Section of Government of India Act*, Government of India, Central Publication Branch, Calcutta.

Crotty, Raymond (1980) *Cattle, Economics and Development*; Commonwealth Agricultural Bureau.

Crotty, Raymond (1983) EEC Interest in Operation Flood, *Economic and Political Weekly*, 2nd April 1983.

Curth, Louise Hill (2002) The Case of the Brute Beast: Animals and the Seventeenth Century Animal Market Place, *Social History of Medicine*, Vol 15, No 3, Pp. 375–392.

Damodaran, Harish (2012) Cow Belt or Buffalo Nation, *The Hindu Business Line*, 18th April 2012. http://www.thehindubusinessline.com/opinion/columns/harish-damodaran/article3328532.ece

Damodaran, Harish (2012) Amul's Gen-Next farmers, Business Line, 20th November, 2012. http://www.thehindubusinessline.com/opinion/columns/harish-damodaran/amuls-gennext-farmers/article4115993.ece

Damodaran, Harish (2015) The Big Picture: A new churning, The Indian Express, 25th January 2015. http://indianexpress.com/article/india/india-others/the-big-picture-a-new-churning/99/

Dandekar, V. M. (1964) Problems of Numbers in Cattle Development, *The Economic Weekly*, Annual Number, (February 1964), pp. 351–355. https://www.epw.in/system/files/pdf/1964_16/5-6-7/problem_of_numbers_in_cattle_development.pdf

Das, K Madhava (1960) *E. M. Hough's Co-operative Movement in India*, Oxford University Press, London.

Dasgupta, Satish Chandra (1945) *The Cow in India*, Khadi Pratisthan, Calcutta.

Datta, T. N. and B. K. Ganguly (2002) Analysis of Consumer Expenditure Pattern in States with Special Reference to Milk and Milk Products, National Information Network, National Dairy Development Board.

Dharampal (2002) *The British Origin of Cow Slaughter in India: With Some British Documents on the Anti-Kine Killing Movement, 1880–1894*, Society for Integrated Development of the Himalayas, Massoorie.

Dharmadhikary, Shripad et al (2005) *Unravelling Bhakra: Assessing the Temple of Resurgent India*, Manthan Adhyayan Kendra, Badawani, pp. 57–88.

Dogra, Bharat (1982) Milk Production: Not Such a Flood, *Economic and Political Weekly*, (May 15 1982).

Dorsten, F. Van (1990) The Impact of AMUL on the Milk Economy of the Kheda District, in Doornbos, Martin and K. N. Nair (eds) *Resources, Institutions and Strategies: Operation Flood and Indian Dairying*, Sage Publications, New Delhi.

Duckham, A. N. (1932) *Animal Industry in British Empire*, Oxford University Press, London.

E. B. J. (1896) *Cows in India and Poultry: Their Care and Management*.

Editorial (1955) *The Economic Weekly*, 29th October 1955.

Editorial (1960) *The Economic Weekly*, 29th October 1960.

Editorial (1975) Dairy Development: On the Wrong Tracks, *Economic and Political Weekly*.

Editorial (1978) Operation Flood: Time for a Closer Look, *Economic and Political Weekly*, July 15, 1978; p. 1122–1123.

Ensminger, E. S. (1956) *Research in Animal Husbandry: A Review of Work Done During 1929–1954*, Indian Council of Agricultural Research, Delhi.

Gadgil D. R. and V. R. Gadgil (1940) *Survey of the Wai Taluka*, Gokhale Institute of Politics and Economics, Poona.

Gandhi, M. K. (1949) *Food Shortage and Agriculture*, Navjivan Publishing, Ahmedabad.

Gautam; Dalal, R. S. and V. Pathak (2010) Indian Dairy Sector: Time to Revisit Operation Flood, *Livestock Science*, Vol 127, No 2–3, Pp. 164–175.

Ghare, B. K. (1928) *Manual of Dairy Farming for Indian Students*, Macmillan, Bombay.

Ghose, Sanjay (1994) It's a Cow's Life, *Down To Earth*, (April 15, 1994). http://www.downtoearth.org.in/node/29681

Ghotge, Nitya (2004) *Livestock and Livelihoods: The Indian Context*, Foundation Books, New Delhi.

Grewal, S. S. and P. S. Rangi (1980) Economics and Employment of Dairying in the Punjab, *Indian Journal of Agricultural Economics*, Vol 30, No 4, (October-December), pp. 120–125.

Hanumantha Rao, C. H. (1988) Agricultural Development and Ecological Degradation: An Analytical Framework, *Economic and Political Weekly*, Vol 23, No 52/53, pp. A-142–A-146.

Hardiman, David (1981) *Peasant Nationalists of Kheda*, Oxford University Press, Delhi.

Harris, Marvin (1966) The Cultural Ecology of India's Sacred Cattle, *Current Anthropology*, Vol VII, No 1, pp. 51–59.

Heredia, Ruth (1997) *The Amul India Story*, Tata McGraw-Hill, New Delhi.

Heston, Alan (1971) An Approach to the Sacred Cow of India, Current Anthropology, Vol 12, pp. 191–197.

Hope, Victor (1937) Opening Remarks at Second Cattle Conference, *Proceeding of the Cattle Conference held at Shimla on 25th May 1937*, Government of India Press, Shimla.

Hough, Eleanor M (1932) *The Cooperative Movement in India*, P. S. King & Son, Westminster, London.

Howard, Albert and Gabrielle L C Howard (1929) *The Development of Indian Agriculture*, Oxford University Press, Bombay.

Huria, Vinod and K T Acharya (1980) Dairy Development in India: Some Critical Issues, *Economic and Political Weekly*, Vol 15, No 45–46, pp. 1933–1942.

Jalan, B. L. (1951) *Food Problem in India*, Marwari Chamber of Commerce, Calcutta.

Joshi, Lemuel Lucas (1915) Bacteriological Examination of the Bombay Milk Supply and the Question of Bacterial Standards for India, *The Journal of Dairying and Dairy Farming in India*, Vol III, Part 1, pp. 05–36.

Joslen, F. (1905) *Cattle of the Bombay Presidency*, Bulletin No 26, Department of Land Revenue and Agriculture, Bombay.

Kahlon, A. S. and D. S. Gill (March 1967) *A Case for Mechanising Selected Agricultural Operations in Punjab*, Directorate of Economics and Statistics, Ministry of Food and Agriculture, Government of India.

Kamath, M V (1996) *Milkman from Anand: The Story of Varghese Kurien*, Konark Publishers, New Delhi.

Kapur, Depinder et al (2012) *A Commons Story: In the Rain Shadow of Green Revolution*, Foundation for Ecological Security, Anand. https://www.nabard.org/Publication/OP_60_commons_story.pdf

Katariya, M. S. (1982) *Geography of Indian Livestock and White Revolution*, Vishal Publications, Kurukshetra.

Keen, Bernard Augustus (1932) 'The Real Problem in India', Paper read at the Royal Institute of International Affairs on 1st March 1932. Quoted in Wright, Norman C. (1937) *Report on the Development of Cattle and Dairy Industries of India*, Government of India Press, Shimla. p. 50.

Khurana, Indira (1997) Vanishing Breeds, *Down To Earth*, September 15, 1997. https://www.downtoearth.org.in/coverage/vanishing-breeds-24429

Khurody, Dara N. (1974) *Dairying in India: A Review*, Asia Publishing House, New Delhi.

Kolhatkar, V. Y. (1946) *Reconstruction of Indian Agriculture*, Popular Book Depot, Bombay.

Knight, Henry (1954) *Food Administration in India 1939-1947*, Stanford University Press, Stanford.

Kurien, Varghese (1983) Letter to Editor, *Economic and Political Weekly*, (April 02, 1983).

Kurien, Varghese (as told to Gauri Salvi) (2005) *I too had a Dream*, Roli Books, New Delhi.

Lal, Gopal (1948) The Lucknow Cooperative Milk Supply Union, *The India Cooperative Review*, Vol XIV, No 03, July–September 1948, pp. 168–174.

Lensch, J. (1987) *Problems and Prospects of Cattle and Buffalo Husbandry in India*, Hamburg, Germany.

Low, C. E. (1912) The Supply of Agricultural Cattle in India, *The Agricultural Journal of India*, Vol VII, Part IV, pp. 331–342.

Madan, G. R. (2007) *Cooperative Movement in India*, Mittal Publications, New Delhi.

Majumdar, N. A. (1960) Cow Dung as Manure, *The Economic Weekly*, May 14, pp. 713–714.

Mallik, S. N. (1929) Review of L. L. Sundara Rama's *Cow Protection in India*, in *Asiatic Review*, XXV, (January 1929), pp. 161–162.

Matson, J. (1933) Inefficiency of Cattle in India through Disease, Agriculture and Livestock in India No 1, pp. 227–228.

Mazoomdar, Jay (2013) Desi Cow, Almost Extinct, *Tehelka*, February 2013.

Mishra, S. N. (1973) Surplus Cattle in India: A Critical Study, *Sociological Bulletin*, No 22, Pp. 297–308.

Mishra, S. N. (1978) *Livestock Planning in India*, Vikas, New Delhi.

Mishra, Saurabh (2011) Beasts, Murrains and the British Raj: Reassessing Colonial Medicine in India from the Veterinary Perspective, 1860–1900; *Bulletin of History of Medicine*, Vol 85, No 4, Winter 2011, pp. 587–619.

Mishra, Saurabh (2014) 'Introduction: Veterinary History Comes of Age', in a virtual issue of *Social History of Medicine* on the theme of Veterinary History (September 2014). http://www.oxfordjournals.org/our_journals/sochis/veterinaryhistoryintro.pdf

Mishra, Saurabh (2015) *Beastly Encounters of the Raj: Livelihoods, Livestock and Veterinary Health in North India, 1790-1920*, Manchester University Press, Manchester.

Mollison, J. and L. French (1907) The Montegomary and Sind Breeds of Cattle, *Agricultural Journal of India*, Vol II, No 2, p. 255.

Moore, John (1921) *Army Veterinary Service in War*, H. & W. Brown, London.

Morgan, J. W. A. (1907) The Government Cattle Farm, Hissar, Punjab, *The Agricultural Journal of India*, Vol II, Part IV, Pp. 364–373.

Mukherjee, B. B. (1937) *Agricultural Marketing in India*, Thacker Spink & Co Ltd., Calcutta.

Mukherjee, Radhakamal (1938) *Food Planning for Four Hundred Million*, Macmillan, London, p. 138; Cited in Raj, K. N. (1971) 'India's Sacred Cattle: Theories and Empirical Findings', *Economic and Political Weekly*, Vol 6, No 13 (27th March 1971), Pp. 717+719–722.

Naidu, K. Muthukrishna (1948) The Madras Cooperative Milk Supply Union, *The India Cooperative Review*, Vol XIV, No 03, July-September 1948, pp. 175–180.

Nair, K N (1980) Operation Flood: Some Constraints and Implications, *Economic and Political Weekly*, February 23, 1980.

Narain, Sunita (2019) India's Cow Crisis Part 3: Brutal to Kill India's Ancient Uber Economy, *Down To Earth*, (8th January 2019). https://www.downtoearth.org.in/news/agriculture/india-s-cow-crisis-part-3-brutal-to-kill-india-s-ancient-uber-economy-62752

Nath, V. (1965) The New Village II: Technological Change in Agriculture, *Economic and Political Weekly*, April 14, 1965. Pp. 713–722.

Niyogi, J. P. (1940) *The Cooperative Movement in Bengal*, Macmillan.

O'Gorman, P. W. (1895) Dairy Produce in India, A paper read at the Indian Medical Congress, 1894; Caledonian Printing Works, Calcutta.

Olver, Arthur and M. Vaidyanathan (1933) Assessment of the Annual Contribution of Livestock to Indian Economy, in *Proceedings of the First meeting of Animal Husbandry Wing of Board of Agriculture in India*, Government of India Press, Shimla.

Olver, Arthur (1942) Animal Husbandry in India, *Journal of Royal Society of Arts*, Vol 90, No 4614, Pp. 431–451.

Onselen, C. Van (1972) Reactions to Rinderpest in South Africa, 1896–97, *The Journal of African History*, Vol 13, No 3, pp. 473–488.

Parthsarathy, G. (1991) White Revolution, Dairy Cooperatives and Weaker Sections, *Economic and Political Weekly*, Vol 26, No 52 (December 28, 1991).

Patel, Bhailal (1996, second edition) *Bhaikaka naa SansmaraNo*, Sastu Sahitya Vardhak Karyalay, Ahmedabad.

Patel, Sujata (1990) Anand Model in Doornboss, Martin and K. N. Nair (eds) *Resources, Institutions and Strategies: Dairy Aid and Operation Flood*, Sage Books, New Delhi.

Patel, M. D. and C. N. Dave (1944) Kankrej Husbandsmen of Gujarat, *Indian Farming*, Vol V, No 5, (July) Pp. 276–278.

Patra, Bijoy Basant (1994) Milking the Cattle, *Down To Earth*, (May 31, 1994). http://www.downtoearth.org.in/node/31995

Pereira, Charles (1982) Bernard August Keen, (5 September 1890 – 5 August 1981), *Biographical Memoirs of the Fellows of Royal Society*, Vol 28, Pp. 205–223.

Perkins, John H. (1997) *Geopolitics and the Green Revolution: Wheat, Genes and the Cold War*, Oxford University Press, New York.

Philips, R. W. (1945) Impressions of India, *The Scientific Monthly*, Vol 61, No 5, (November 1945), pp. 397–412.

Prasad, Archana (2006) Review of Ecology, Colonialism and Cattle, in Social Scientist, Vol 34, Nos 11–12, Pp. 80–82.

Puniyani, Ram (2017) Hindutva and Cow Agenda: Love for Cow or Hatred for "Others"? Mainstream, Vol LV, No 30 (July 15, 2017). https://www.mainstreamweekly.net/article7308.html

Queripel, A. E. (1879) Cattle plague in India, *The Veterinary Journal and Annals of Comparative Pathology*, Vol IX, Issue 11, November 1879, Pp. 310–316.

Rajendran, K. and Samarendu Mohanty (2004) Dairy Cooperatives and Milk Marketing in India, *Journal of Food Distribution Research*, July 2004, Pp. 34–41.

Ramdas, Sagari (2015) Death of Small-Farmer Dairies amidst India's Dairy Boom, *Economic and Political Weekly*, Vol L, No 19, pp. 21–23.

Ramakrishnan, K. C. and V. S. Narasimhan (1948) The Coimbatore Milk Supply Union, *The India Cooperative Review,* Vol XIV, No 03, July–September 1948, pp. 181–185.

Ramaswamy, N S (1983) Draught Animal Power, *India International Centre Quarterly*, Vol 10, No 3, pp. 309–320.

Ramaswamy, N. S. (1985) *Draught Animal Power: Socioeconomic Factors*, in Copland, J. W. (ed) *Draught Animal Power for Production*, ACIAR Proceedings, ACIAR, Canberra, Australia, pp. 20–25.

Randhawa, M. S. (1962) *Agriculture and Animal Husbandry in India*, Indian Council of Agricultural Research, New Delhi.

Randhawa, M. S. (1986) *A History of Agriculture in India*, Vol IV (1947–1981), ICAR, Pusa, Delhi.

Ranjhan, S. K. (1974) *Feeding of Cattle, Sheep and Goats*, Animal Nutrition Division, IVRI, Izzatnagar.

Rao, G. A. (1956) Milk for Millions, *Economic and Political Weekly*, (May 19), Pp. 585–587.

Rao, Venkatsubramaniam and Wit (1995) Consequences of crossbreeding of Cattle, *Economic and Political Weekly*.

Ratnagar, Shereen (2004) *The Other Indians: Essays on Pastoralists and Prehistoric Tribal People*, Three Essays Collective, New Delhi.

Ratnagar, Shereen (1991) Pastoralism as an Issue in Historical Research, *Studies in History*, Vol 7, No 2, Pp. 181–193.

Rebello, R. J. (1973) *The Organisation of Government of Mysore*, Indian Institute of Public Administration, Mysore Regional Branch, Bangalore.

Read, W. S. (1937) *The Government Cattle Farm, Hissar: A History and Guide*, Superintende, Government Printing, Punjab, Lahore.

Rosen, George (1985) *Western Economists and Eastern Societies*, Oxford University Press, Bombay.

Russell, E. John (1937) *Report on the Work of the Imperial Council of Agricultural Research in Applying Science to Crop Production in India*, Government of India Press, Shimla.

Samanta, Samiparna (2016) Cattle, Cruelty, Cow Doctors: Examining Animal Health in Rural Bengal, 1850-1920, in Kumar, Deepak and Bipasha Raha (eds) (2016) *Tilling the Land: Agricultural Knowledge and Practices in Colonial India*, Primus Books, Delhi. Pp. 214–237.

Samanta, Samiparna (2021) *Meat, Mercy and Morality: Animals and Humanitarianism in Colonial Bengal, 1850–1920*, Oxford University Press, New Delhi.

Schneider, Harold K. (1983) Review of *Cattle, Economics and Development* by Raymond Crotty, *Economic Development and Cultural Change*, Vol 32, No 1, (October 1983), Pp. 222–226.

Sen, K. C. and S. C. Ray (1941) Animal Husbandry and Crop Planning in India, *Science and Culture*, Vol VI, No 2, (June 1941), Pp. 684–689.

Shah, K. T. (ed) (1948) *Report of Sub-committee on Animal Husbandry and Dairying*, National Planning Committee Series, Vora and Co Publishers Ltd. Bombay.

Shah, Narsimh M. (1940) *Dudh: Sarva SampurN Khorak* (in Gujarati), Gujarat Vernacular Society, Ahmedabad.

Shah, Tushar (1990) 'Unconvincing Critique of Operation Flood,' *Economic and Political Weekly*, Vol XXV, No 34 (May 26, 1990)

Shastri, C. P. (1960) 'Bullock Labour Utilisation in Agriculture,' *The Economic Weekly*, October 29, pp. 1585–1592.

Singh, K (1995) Research Priorities in Improving Animal Agriculture by Agro-Ecological Zones in India, in Devendra C., Gardiner, P. (Eds.) *Global Agenda for Livestock Research: Consultation for South Asia Regions*, ICRISAT Asia Centre, Patancheru, 6–8 June 1995, ILRI, Nairobi, Kenya, pp. 85–101.

Singh, Narendra (1979) Unloading White Revolution, *Economic and Political Weekly*, (May 19, 1979), Pp. 874–875.

Singh, Surendar (1979) Operation Flood II: Some Constraints and Implications, *Economic and Political Weekly*, October 27, 1979.

Singh, Surenddar and R K Sharma (1981) Some Implications of Area Choice in Operation Flood II Programme, *Economic and Political Weekly*, March.

Singh, S. P. and P. L. Kelly (1981) *AMUL: An Experiment in Rural Economic Development*, Macmillan, New Delhi.

Sivasundaram, Sujit (2005) Trading Knowledge: The East India Company's Elephants in India and Britain, *The Historical Journal*, Vol 48, no 1, pp. 27–63.

Smith, Frederick (1927) *A History of Royal Army Veterinary Corps* (1796–1919), Bailier, Tindall & Cox, London.

Smith, William (1927) Cattle Improvement Cooperative Scheme, *Young India* (04-08-1927), Navjivan Press, Ahmedabad.

Sood, Jyotika (2014) New Milky Way, *Down To Earth*, 15th February 2014. http://www.downtoearth.org.in/content/new-milky-way

Stow, A. M. (1910) *Cattle and Dairying in Punjab*, Government of India Press, Shimla.

Streeter, C. P. (1969) *A Partnership to Improve Food Production in India*, The Rockefeller Foundation Special Report, pp. 93–99.

Sukhatme, P. V. (1970) Nature of Protein Problem and its Implications for Policy Measures, Paper presented at National Food Congress, New Delhi.

Sundara Ram, L. L. (1926) The Indian Attitude towards Cow Protection, *The Maha-Bodhi*, Vol XXXIV, No 10, (October), pp. 497–506.

Sundara Ram, L. L. (1927) *Cow – Protection in India*, The South Indian Humanitarian League, Madras.

Swaminathan, Madhura & Yoshifumi Usami (2016) Women's Role in Livestock Economy, *Review of Agrarian Studies*, Vol 6, No 2, Pp. 123–134. https://ras.org.in/697c22e48e9cb0207d3e5d7560c971f6

Thorner, Daniel and Alice Thorner (1959) Ploughing the Plan Under: From Lathes to Ploughshares, *The Economic Weekly* Republished in *Land and Labour in India* (1962).

Trivedi, A. B. (1949) *Post-War Gujarat: Economic Survey after World War II*, A. B Trivedi, Khalsa College, Bombay.

Verma, A. M.; Singh, P. K. and Pandey, P. (2008) Organic Animal Husbandry in India, *Indian Farming*, July, Pp. 29–30.

Vijapurkar, Mahesh (1986) Commodity Aid for Dairying: Threat that was converted into Opportunity, *The Hindu*, 13th May 1986. http://www.indianet.nl/a860513.html

Voelcker, John Augustus (1893, 1897 2nd ed.) *Report on the Improvement of Indian Agriculture*, Office of the Superintendent of Government Printing, Calcutta.

Walton, H. G. (1903) *A Monograph on Training and Working in Leather in the United Provinces*, Government Press, Allahabad.

Warner, James (1951) *Dairying in India*, Macmillan & Co Ltd., Calcutta.

Whyte, Robert O. (1958) International Development of Grazing and Fodder Resources: Part V India, Grass and Forage Science, Vol 13, Issue 3, Pp. 203–212.

Whyte, Robert O. (1961) *Grasslands and Fodder Resources of India*, Indian Council of Agricultural Research, New Delhi.

Whyte, Robert O. (1967) *Milk Production in Developing Countries*, Faber and Faber, London.

Whyte, Robert O. (1968) *Land, Livestock and Nutrition*, Commonwealth Agriculture Bureau.

Woodroffe, John (1926) 'Introduction' to Nilananda Chatterjee's book, *The Condition of Cattle in India: Being an Inquiry into the Causes of the Present Deterioration of Cattle with Suggestions for their Remedy*, All India Cow Conference Association, Calcutta.

Woodroffe, John (1926) A Note on the Cattle of India: With Special Reference to Breeding, Pasturage, Slaughter and Exports, Reproduced in Chappterjee, Nilananda (1926) *The Condition of Cattle in India: Being an Inquiry into the Causes of the Present Deterioration of Cattle with Suggestions for their Remedy*, All India Cow Conference Association, Calcutta, pp. 13–15 in Appendices.

Wright, N. C. (1937) *Report on the Development of the Cattle and Dairy Industries of India*, Government of India Press, Shimla.

Wright, N. C. (1952) Some Problems of Tropical Dairying, *Proceeding of the Royal Society on Dairying*, 8th January 1952, Morning Session, pp. 149–157.

Index

H. Upadhyaya, *Critical Insights on Colonial Modes of Seeing Cattle in India (1850–1980)*, Asia in Transition 27, https://doi.org/10.1007/978-981-97-1560-2

W

SPRINGER NATURE

GPSR Compliance

The European Union's (EU) General Product Safety Regulation (GPSR) is a set of rules that requires consumer products to be safe and our obligations to ensure this.

If you have any concerns about our products, you can contact us on ProductSafety@springernature.com

In case Publisher is established outside the EU, the EU authorized representative is:

Springer Nature Customer Service Center GmbH
Europaplatz 3
69115 Heidelberg, Germany

The manufacturer's authorised representative in the EU is Springer Nature Customer Service Centre GmbH, Europaplatz 3, 69115 Heidelberg, Germany. If you have any concerns regarding our products, please contact ProductSafety@springernature.com

Printed and bound by CPI Group (UK) Ltd, Croydon, CR0 4YY
01/10/2026
02225925-0012